Pure and Modern Milk

Pure and Modern Milk

An Environmental History since 1900

KENDRA SMITH-HOWARD

OXFORD
UNIVERSITY PRESS

OXFORD

UNIVERSITY PRESS

Oxford University Press is a department of the University of
Oxford. It furthers the University's objective of excellence in
research, scholarship, and education by publishing worldwide.

Oxford   New York

Auckland   Cape Town   Dar es Salaam   Hong Kong   Karachi
Kuala Lumpur   Madrid   Melbourne   Mexico City   Nairobi
New Delhi   Shanghai   Taipei   Toronto

With offices in

Argentina   Austria   Brazil   Chile   Czech Republic   France   Greece
Guatemala   Hungary   Italy   Japan   Poland   Portugal   Singapore
South Korea   Switzerland   Thailand   Turkey   Ukraine   Vietnam

Oxford is a registered trade mark of Oxford University Press
in the UK and certain other countries.

Published in the United States of America by
Oxford University Press
198 Madison Avenue, New York, NY 10016

Library of Congress Cataloging-in-Publication Data
Smith-Howard, Kendra.
Pure and modern milk : an environmental history since 1900 /
Kendra Smith-Howard.
    pages   cm
Includes bibliographical references and index.
ISBN 978–0–19–989912–8 (alk. paper)
1. Dairy products industry—United States—History.
2. Milk—Quality—United States—History—20th century.
3. Dairy products—United States.   4. Dairy products—United States—
Marketing.   I. Title.
QP144.M54S65 2013
636.2'142—dc23   2013019901

1 3 5 7 9 8 6 4 2

Printed in the United States of America
on acid-free paper

CONTENTS

ACKNOWLEDGMENTS

It gives me great pleasure to be able to thank all of those who have inspired and supported this project in various stages. Some asked interesting questions, others lent a listening ear. Still others slogged through messy drafts or offered consolation on dreary writing days. Whatever their contribution, I could not have completed this book without them. I only hope that I can be as gracious and generous with my time as they have with theirs.

This project began as a dissertation at the University of Wisconsin-Madison. I arrived at the university feeling very lucky to be able to study with Bill Cronon. The longer I have known Bill, the more deeply I have come to admire his scholarship and his commitment to his students. He is a first-rate mentor. I am so grateful for the many ways he has helped me find my path in the field. Nan Enstad, Jess Gilbert, Judith Walzer Leavitt, and Gregg Mitman made suggestions in my dissertation defense that helped greatly as I revised the manuscript. I appreciate all they taught me about researching and writing history. In coursework and other settings, I benefited greatly from the tough questions and models of scholarship of the entire History Department faculty at the University of Wisconsin. I especially thank Colleen Dunlavy, Steve Kantrowitz, and Al McCoy for engaging and pushing me as a graduate student.

Of all my teachers, my undergraduate advisor, Jim Farrell, is most directly responsible for this book. Without him, I would never have even imagined going to graduate school or becoming a historian. Nor would I have had any confidence to be playful with ideas. Who but Jim would ask an earnest undergraduate, as her first research assignment, to find out how, precisely, shit happens? Blending historical rigor with a wry wit, Jim planted the seeds for this quirky project. He is a treasure to all his students, and deeply missed. Other teachers lit my curiosity and held me to high standards in high school and college; I especially thank William Hohulin, Frieda Knobloch, Judy Kutulas, and Dolores Peters.

Historians owe a great debt to the librarians and archivists who provide access to and insight about the documents and records in their charge. In this, I am no exception. My thanks to Roger Horowitz at the Hagley Museum and Library; Susan Strange, John Flecknor, Kay Peterson, and Fath Davis Ruffins at the Archives Center, National Museum of American History; Janice Goldblum at the Archives of the National Academies; Nelson Lankford at the Virginia Historical Society; John Skarstad at the University of California-Davis; Tara Vose at the Strawbery Banke Museum; and the staffs of the California State Parks Office, Western Historical Manuscripts Collection, Minnesota Historical Society, Montana Historical Society, New York State Library and Archives, Vermont Historical Society, and the National Archives and Records Administration in College Park, Maryland. Special thanks to the following librarians at the University of Wisconsin and Wisconsin Historical Society: Michaela Sullivan-Fowler, at Ebling Library, who offered an endless supply of interesting tidbits; Rick Pifer and Harry Miller, at the historical society's archives reading room, who delivered good cheer with the documents; and the entire staff at Steenbock Library, who retrieved document after document on my behalf without complaint. The Interlibrary Loan Department at the University at Albany made my life infinitely easier by tracking down obscure agricultural bulletins.

I am grateful for the generous support the project received. Early on, a short-term grant-in-aid from the Hagley Museum and Library gave me a sense of the wealth of possibilities available and buffeted my confidence. A semester-long predoctoral fellowship from the Smithsonian Institution's National Museum of American History allowed me to devote myself to this project full time in a city rich with archives. More important, museum curators Pete Daniel and Jeffrey Stine introduced me to a network of people and sources that made this project stronger. The University of Wisconsin-Madison history department provided support through a semester-long writing grant at a critical time. Grants from the Virginia Historical Society and the University at Albany's Faculty Research Award Program allowed the project to take on a national scope. Finally, a research leave sponsored by the New York State United University Professions' Joint Labor-Management Committee allowed me time to integrate new findings and rewrite the manuscript. I am grateful to the United University Professions and to history department chair Richard Hamm and Dean Elga Wulfert of the College of Arts and Sciences for sponsoring my application.

It took scores of drafts and the feedback of many readers to help me make sense of the muddled mess of milky evidence uncovered in the research stage. Their constructive criticism helped a great deal; any remaining errors or omissions are my own. I presented parts of this research at meetings of the Agricultural History Society, Business History Society, and American Society for Environmental History. I thank commentators Kathleen Brosnan, Pete

Daniel, Thomas Pegram, and Amy Slaton for comments at these meetings. I also had the pleasure of participating in stimulating workshops at the University of Western Ontario, Montana State University, University of Georgia, and the University of Guelph. I thank the organizers of these meetings, as well as Shane Hamilton, Alan MacEachern, Sara Pritchard, Paul Sutter, and William Turkel for their advice on these occasions.

On a day-to-day basis, I've been blessed with generous and supportive colleagues. When I began this project, Will Barnett, Jim Feldman, Mike Rawson, and Tom Robertson kindly invited me to join their dissertation writing group and blazed a trail of success ahead of me. I've appreciated their sage advice. Despite reading the crudest versions of these chapters, my dissertation group, and Dawn Biehler, Todd Dresser, Holly Grout, Michel Hogue, Jen Martin, Chris Wells, and Keith Woodhouse never lost faith. Thanks for standing by and waiting patiently for things to take shape. Davis Brown, at the University of Wisconsin Writing Center, was an especially skillful tutor. Countless others offered their friendship in graduate school. Thanks to Ikuko Asaka, Winton Boyd, Scott Burkhardt, Andrew Case, Yosef Djakababa, Sam Graber, Kori Graves, Brian and Diana Hutchinson, Jessica Martin, Story Matkin-Rawn, Neela Nandyal, Abby Neely, Stacey Pelika, and Amrys Williams for making my life so rich.

At the University at Albany, State University of New York, a new cast of characters helped clarify and correct my thinking. Elise Andaya, Elizabeth Berman, Jennifer Dodge, Kristin Hessler, Abby Kinchy, and Barbara Sutton kept the writing momentum going. Jackie Mirandola Mullen offered research support at a critical moment. I am grateful to all my colleagues in the history department for their goodwill and professionalism, but especially to Mitch Aso, Iris Berger, Sheila Curran Bernard, Carl Bon Tempo, Rick Fogarty, Susan Gauss, Dave Hochfelder, Amy Murrell Taylor, and Laura Wittern-Keller for commenting on parts of the manuscript, offering support, and answering questions about the publication process. Richard Hamm earns hearty thanks for his leadership and counsel as department chair. Irene Andrea and Marlene Bauman's efficiency and unfailingly good spirits are heartening.

Susan Ferber, my editor at Oxford University Press, has proven a careful critic and a thoughtful person. Her greatest gift was finding anonymous reviewers who offered perceptive, extensive, and detailed comments on the manuscript. I give my deepest thanks to them. I also appreciate the grace with which she, and the entire staff of Oxford University Press, has guided an at-times-rattled first-time author. I also thank anonymous reviewers from *Agricultural History*, and its editor, Claire Strom, for commenting on a section of chapter 5, which appeared in that journal in a different version in 2010.

Most of all, I am grateful for my family's unwavering support. Although none of my grandparents will read this history, their stories are at its heart. It was summers spent on the Smith family farm and endless conversations at family dinners about weather and corn prices through which I was introduced to rural history. When I began this project, my grandfather L. Dale Beall, a retired dairy farmer, entertained questions about his experiences. I wish he could have read this book. I am grateful, however, for the opportunities we had to talk about cream separators and milking parlors. My grandmother Veleanor Beall's account of her contributions to the farm income convinced me to stay at the computer writing and working longer than I might have otherwise. My mother-in-law, Lucille Howard, has been curious about this project since we met; I am happy to finally deliver something of mine for her reading list. My parents, Kenneth and Lorna Smith, have loved me and supported this work unfailingly. Even when my academic route befuddles them, they come to my aid with understanding and compassion. I am lucky to be their daughter, and I dedicate this book to them.

My biggest debt is to Ken Howard. No one has lived with this project longer or more intimately. He has laughed at endless puns about milk, listened attentively to mind-numbing details about its manufacture, and endured his partner's distraction. He has cheered me in the lowest moments, celebrated small accomplishments, and been a constant source of comfort. Thank you. My little boys, Sebastian and Alden, powered this project to its completion and made me think about milk in new ways. I look forward to new adventures and ice cream cones with them by my side.

Pure and Modern Milk

Introduction

On November 20, 1967, Philadelphia resident Edna Irwin wrote to the President's Special Assistant on Consumer Affairs, Betty Furness, with a question inspired by the label on her milk carton. "Would you please tell me what the word 'manufactured' means in the processing of milk? If milk is a product of cows, how can humans manufacture it?"[1] Irwin's questions cut to the heart of a paradox that this book explores. From the turn of the twentieth century, milk and dairy foods have simultaneously been presented to consumers as pure products of nature *and* as products remade by human intervention and modern technologies. To make milk readily available as a staple food, dairy farm families, health officials, and food manufacturers have simultaneously stoked human desires for an all-natural product and intervened to ensure milk's safety and profitability.

Glance at the cartons in the dairy aisle of the nearest supermarket and you will encounter the same phenomenon that perplexed Edna Irwin. On the milk label peddled by Stonyfield Farms, three cows graze on a verdant pasture, backed by a deciduous forest and an undulating hillside. On another, a cow's wide-eyed face, graced with a wreath of meadow flowers, beckons the thirsty drinker. Store-brand dairies festoon their milk bottles with images of red and green barns at sunrise to proclaim the product's wholesomeness.[2] However, the same labels in the dairy case that flaunt meadow flowers and red barns betray a different history, one of human manipulation of milk between farm and supermarket. Words on the carton indicate that milk is "Grade A," "pasteurized," "homogenized," and "vitamin fortified." Multicolored plastic caps help purchasers distinguish between whole, 2 percent, 1 percent, and skim milk, while the nutritional fact panel calculates the meaning of such designations in fat grams and calorie content. The cartons carry expiration dates and advise consumers to keep the product refrigerated. These adjectives and numbers convey a different reality than images of happy cows: harnessing cows' lactation cycles and preparing milk for sale require an extraordinary amount of human intervention. On behalf of pure and plentiful milk, Americans have become as reliant on inspectors to monitor cows for diseases and suppliers to keep milk cool as on idyllic agricultural landscapes.

Though often conceived of as a pure product of nature, milk's nature must be perfected for it to become a healthful human food.

While many Americans buy milk without giving it a second thought, an increasing number approach the act of purchasing food with trepidation. As they assess package labels, consumers consult their taste buds and pocketbooks and consider their nutritional requirements and moral commitments. Despite straightforward recommendations from nutritionists, advertisers, and food reformers, decisions about what to eat are rarely simple. What one eats has become a battleground for environmental politics and the fight against childhood obesity. Perplexed shoppers wonder whether the nutritional or ecological benefits of organic milk justify its premium price. They deliberate whether milk from a local dairy will be fresher or spoil more quickly than ultra-pasteurized milk shipped from a greater distance. They puzzle over how a product sold to add creaminess to coffee can be billed as "fat free." The paradoxes facing consumers in the dairy aisle point to the broader complexities of seeking health and maintaining a sustainable relationship with nature in modern America.

Milk is not the only food long lauded for its natural origins. Nor is it the only food that reaches the marketplace in an altogether different state from that in which it originated.[3] But no other food has so stolidly symbolized natural purity, while simultaneously undergoing dramatic transformations to its material form. How and why has milk been conceptualized as wholly natural, even as it has been churned into manufactured foods like butter and ice cream, and incorporated into products as artificial as Cheez Whiz and wood glue? What ideas and values drove the modification of milk so that it became a staple food for Americans? How have consumers' changing expectations for milk affected the dairy farmers, cows, and rural landscapes central to milk production? This book explores these questions, connecting the development of dairy farming and changing practices of buying milk products from the turn of the twentieth century to the present. It traces the biological and economic processes through which milk has been produced and consumed, and it chronicles the meanings people made from those processes through the stories of four different dairy products: fluid milk, butter, ice cream, and the leftover waste of dairy processing (whey, skim milk, and milk proteins).

Two forces were especially important in making milk, paradoxically, a modified "natural" food: the food's role in the diet of children and its capacity to be transformed into a broad array of consumer products. Milk's status as a staple food for American children stimulated advertisers to stress its "natural" origins and also to modify its biological form. At the turn of the twentieth century, when infant formulas were commonly designated as "artificial" foods, calling cows' milk "natural" helped assuage parents' concerns about the shift from breast- to bottle-feeding. At the same time, the material properties of cows' milk and the biology of cows' bodies defied its safety and availability for infant feeding. One challenge was the seasonality of milk production. Most cows calved in the spring and thus produced more milk in the spring and summer months than in the fall and winter. Cows needed adequate feed to keep lactating, and as pastures dried up in the fall, their milk production dropped off. Another feature of milk that posed a challenge to its use was its propensity to spoil. If left unrefrigerated, milk sours within forty-eight hours of leaving the cow. Getting fresh milk to faraway residents, then, required sophisticated systems of cooling and quick transportation, in an age before electric refrigerators and automobiles. A final challenge to the use of milk as a food for the masses was that the fluid could serve as a medium of deadly communicable diseases, including typhoid, scarlet fever, and bovine tuberculosis. Children's vulnerability to milk-borne diseases drove efforts to reform the substance of milk, remake cows' bodies, and restructure the landscapes from which it came.

If milk's physical properties posed challenges, they also promised possibilities. Transforming milk into other products allowed farm families and manufacturers to transcend some of the natural obstacles that limited the sale of fluid milk, such as spoilage or seasonality. Few foods were so innately well suited to morph into other products as was milk. After being drawn from the cow, milk could be whipped into ice cream, churned into butter, coagulated into cheese, incorporated into candies and breads, or fed to livestock and reach the market as meat. In time, chemists found ways to alter milk even more dramatically, powdering it for long storage, weaving its proteins into cloth fibers, and channeling its sugars into the manufacture of penicillin. Just as milk drinkers needed inspectors and milk companies to guard milk's safety, so too did farm families come to rely upon these manufacturers to transform the raw materials of their farm into salable goods.

Even before milk left the countryside, farm people manipulated its nature, altering cows' bodies and the farm landscape to maximize its production. By bringing greater regularity to cows' feeding and breeding schedules, for instance, farm families could boost cows' capacity for lactation. Supplementing bovine diets with silage or fodder crops helped sustain milk production through the fall and winter months. Mating cows with promising udders to potent bulls increased the likelihood of young calves that would produce record quantities of milk. By modernizing their farms, farm people aimed not just for profits, but also to replace some of the most toilsome farm tasks and to bring order to the unpredictable vagaries of nature, like drought, disease, and insect infestation. In the postwar era, they turned to antibiotics to treat infectious diseases in cows' udders and sprayed pesticides to keep biting flies from irritating their animals. As farm people tweaked the processes of bovine reproduction and lactation, a biological process devised to provide sustenance for young calves simultaneously became bound to a cultural process designed to supply human markets and improve rural comforts.

This book is not the first to discuss the transformation of milk and dairy foods in the twentieth century.[4] What this book contends, however, is that the twentieth-century transformation of milk required not simply changes to the food itself, but also to the farms from which milk came. Growing out of works in environmental history, this book links consumption and production, helping to explain how changing consumer practices and retail techniques changed rural nature.[5] It examines the passage of new public health laws to improve urban milk safety and also traces the implications of these laws for the human and animal inhabitants of the farm. It details the development of new dairying technologies on farm practices, and the challenges such technologies posed to health officials charged with maintaining a safe and plentiful food supply.

Environmental historians have done much to elucidate the relationship between production and consumption. But too often efforts by environmentalists and environmental historians to reconnect consumers to the places from which their food came position the farm as a counterpoint to a rapidly industrializing urban America. Such histories and appeals, like the imagery on dairy labels, encourage consumers to associate milk with a timeless idyllic countryside.[6] This romanticization of rural nature makes it easy to overlook that rural places experienced processes of industrialization in tandem with urban ones in the twentieth century. Modernizing milk was not simply a process that took place once the white beverage reached the city; rural and urban people alike transformed it.

When historians narrate the ways that farm people specialized and adopted capital-intensive methods to increase production, many depict these processes of industrialization as the driver of environmental ills or as a force that erodes authentic rural communities.[7] Such accounts make it difficult to explain why farm people willingly embraced industrialized agriculture. Declensionist narratives tend to portray agricultural modernization as a force that corrupts natural purity, but in some cases, changing nature to be more artificial helped make milk *more* safe and pure. Farm people often had compelling reasons to modernize their operations in ways that dramatically altered nature. By breeding cows artificially, for instance, farm families reduced the real risk of being gored by a bull. Farm people did not uncritically champion all industrial solutions; many sought to adopt new technologies on their own terms. Paying attention to the ways in which they evaluated and understood rural industrialization as it unfolded provides a clearer picture of the human interests served by the manipulation of natural organisms in working landscapes.

To explain why and how dairy farm families understood and reacted to the processes of rural industrialization and assessed its consequences, this book incorporates evidence from farm diaries and records housed in state archives in each of the country's well-established dairy regions: the upper Midwest, New York and New England, and California. But because dairy farms dotted the landscape throughout the nation, and because the problems facing farmers differed by locality, the book also draws on sources gleaned at state archives outside these regions, such as Virginia and Montana, and from interviews of dairy farm families profiled by the Southern Agriculture Oral History Project. Seen together, the archival records illuminate farm practices on a wide variety of operations from small-scale cream producers to large-scale specialized dairies.

Despite the ubiquity of dairy farmers and the admirable efforts of archivists to preserve their records, many state archives hold only a handful of farm diaries that discuss dairy production. Even states with rich agricultural collections tend to focus on the most well-heeled or politically active farm people. I aimed to draw out the experiences of other kinds of farmers into the story with other sources, but the perspectives of specialized and successful farmers appear more prominently than those who struggled. To round out the economic, technological, and environmental trends in the industry, I have consulted the records of milk and agricultural regulators, agriculture experiment station reports, industry records—such as the records of the Badger Cooperative Creamery Company and national trade magazines, *Hoard's Dairyman* and *Creamery and Milk Plant Monthly*.

One of this book's aims is to catalogue the environmental history of rural industrialization. A second is to explain the ways that milk, dairy foods, and the cows and farm landscape from which they came involved a delicate interweaving of human technologies with elements of the nonhuman environment. That Rita Irwin had difficulty understanding whether milk was the product of cows or of humans was for good reason: milk and the cows that produced it were a hybrid of nature and culture. They were products of economic and technological innovations, cultural attitudes, human and animal labor, and environmental forces and structures—soil, plants, water, sunshine, and air.

In this, milk, the cow, and the dairy farm were not unique. Over the past forty years, environmental historians have explicated the processes by which spaces have been marked by human intervention and yet maintain natural qualities.[8] Richard White's *Organic Machine* captures this idea powerfully, revealing how even as people modified the Columbia River to fish and to generate electric power, the river's flow and salmon's migratory journeys remained formidable forces.[9] Other historians challenge perceptions of the city as apart from nature, demonstrating the centrality of natural resource flows to their development and cataloguing the geomorphological transformations of the very ground and water on which urban environments stand.[10] Some historians have even taken environmental history indoors, to examine the natural histories of the factory floor and office space.[11] Together, these works make visible and render significant human-environment interactions not simply in seemingly pristine wild places, but in all spaces in which people live, work, and play. Further, many of these works challenge interpretive frameworks once popular in histories of the environment and technology, depicting technological transformation neither as a fall from wilderness nor as a symbol of humanity's mastery over nature.

More recently, historians have begun to look closely at the blended histories of nature and technology in rural and agricultural spaces.[12] The most provocative contribution of these studies is the idea that nature is not wholly apart from technology, but that nature itself constituted technology, particularly as humans manipulated the bodies of animals and plants. Thus, historian Edmund Russell has urged historians to recast Leo Marx's idea of the "machine" intruding upon "the garden" and instead to explain how the garden (nature) formed the machine (technology).[13] This book takes up Russell's call, paying attention to the ways that the farm landscape, cows' bodies, and dairy foods were standardized and modified by human action, and yet their natural attributes remained critical to dairy production. Whether through the stories of bacterial cultures used to flavor butter, the cement milk tank whose cooling waters were pumped up by a windmill, or the sturdy, barrel-bodied cow being led to a breeding stall, readers will encounter a host of examples of the ways that the material environment acted and, combined with human intentions, produced forms difficult to classify as either creatures of nature or artifacts of culture.

Furthermore, this book explores the ways Americans conceptualized the relationship between nature and technology by calling attention to the central word used to describe milk: purity. No concept was more important in capturing the mix of biological and human processes necessary to make milk a safe and viable human food than this one. Americans' ideals about milk purity shifted over time, in tandem with their changing ideas about nature and modernity. Over the twentieth century, Americans identified a different role for nature in purifying milk. At the turn of the century, what imperiled milk's purity was its nature. Its perishability made it difficult to transport. Its tendency to spoil and carry bacteria threatened those who drank it with digestive and communicable diseases. As pasteurization and refrigeration minimized these risks, the perceived threats to milk shifted from elements of nature, such as bacteria, flies, and spoilage to human technologies, such as pesticide residues and radioactive particles. Nature, once conceived of as the primary threat to milk's purity, was envisioned by postwar Americans as the food's primary source of purity. Ironically, as Americans revised what constituted "pure" milk, the very technologies that once promised to protect milk from the hazards of nature, such as antibiotics or pesticides, became threats to milk's purity themselves.

Americans' turn to nature as the source for milk's purity was itself a modern phenomenon. Only when consumers came to believe that the food on which their lives depended was somehow unnatural and alienating could Americans seek to get back to nature through their diets. The kind of milk Americans put on their tables reflected their cultural expectations for purity and convenience as much as their physical needs for sustenance.

Finally, this book investigates the ways in which changing consumer practices and consumer culture transformed the physical spaces of dairy farms and perceptions about them. Although the field of environmental history has traditionally done more to elucidate how the activities of economic producers (like farmers, anglers, or miners) remake the landscape than those of consumers, in recent years, environmental historians have begun to think more about the effects of consumer behavior on the environment.[14] These creative studies have charted the changing economic and cultural processes through which plants and animals became desired commodities, and documented the ways that the food industry and advertisers transformed consumers' views of nature.[15]

Although concerns about "the consumer" came to have increasing influence in twentieth-century politics, and the collective actions of consumers fundamentally altered the physical environment, the actions of individual consumers can be difficult to track.[16] To get at how and why consumers purchased dairy foods and the meanings they made from those transactions, I have relied upon the papers of consumer organizations, government bodies, surveys of consumer behavior conducted by dairy organizations, cost-of-living surveys, and women's magazines and advertisements. It is easier to read from these documents what health officials and dairy manufacturers believed consumers needed and desired than how consumers expressed these needs and desires themselves. My aim in this book, then, is less to uncover the motivations of individual purchasers of milk, butter, ice cream, or cheese than it is to explain physical settings, economic structures, and political mechanisms through which those purchases took place and became meaningful. As historian Ruth Oldenziel explains, industry, state agencies, trade unions, and other groups mediated the ways that consumers exerted power in the marketplace.[17]

New practices of food retail and distribution altered the way that dairy consumers and producers made decisions. Over the course of the twentieth century, strategies of mass retailing, state policies, and expert recommendations came to play a greater role in shaping actions on the dairy farm and in the grocery aisle alike. The development of chain stores, and later supermarket retailers, altered the very form by which foods like butter and ice cream reached consumers and encouraged the development of quality standards on the farm. State agencies defined milk purity and legitimized some dietary practices in ways that privileged some foods and farming practices and denounced others. Farm people and consumers often reacted to these processes. Farmers who relied heavily on butter sales protested when government nutritionists presented margarine as a nutritious food. Consumers argued for more detailed labels on ice cream and other manufactured dairy foods. That the president had a Special Assistant on Consumer Affairs to whom Edna Irwin could address her puzzlement over milk is indicative of these broader changes.

The structures and institutions of consumer society did not simply alter how Americans understood food; they also reframed how consumers came to know nature. As women shifted from breast to bottle, the act of ingesting milk remained an embodied experience. People tasted the cool, creamy fluid on their tongues and sniffed off-flavors in a bottle that had soured. But with the development of mass consumer society and scientific theories about milk's role in disease transmission, consumers began to consider elements difficult to ascertain from merely seeing, tasting, or smelling, such as bacterial counts, vitamin content, or parts per million of pesticide residues, as they assessed milk's healthfulness. After World War II, when concerns about milk safety began to focus on the bioaccumulation of pesticide residues and strontium-90, farm families altered cows' diets and crop-raising practices to minimize cows' exposure. Employees of state and municipal health departments and federal agencies like the U.S. Department of Agriculture (USDA) and the Food and Drug Administration (FDA) took on an ever-more important role in reassuring milk drinkers of the product's freshness and safety. Herein lay the impetus for supermarket milk cartons laden with nutritional facts and words like "Grade A pasteurized" inscribed on their labels.[18]

Over the course of the twentieth century, consumers revised their vision of the appropriate way to achieve milk purity. In the Progressive Era, consumers insisted on the establishment of local, state, and federal standards for food purity. Purchasers of dairy foods became ever-more reliant on the state as arbiter of food's wholesomeness during the interwar period, pushing state agencies to guard the safety of manufactured dairy foods as well as fluid milk. By the 1920s and 1930s, consumers assessed the labels, looking favorably on phrases like "USDA approved." After World War II, consumers increasingly relied on federal standards crafted in Congress and upheld by the FDA, to ensure milk's purity as it crossed state lines. Since the 1970s, consumers' trust in the state as an arbiter of food safety has declined. Still seeing themselves as part of a "consumer movement," many milk and dairy purchasers in recent times reject state inspection and seek to monitor food safety themselves. Consumers' push for transparent measures of food's contents and for a role in defining food safety constituted an important way of adjusting to the products of modern agriculture.

To make milk pure and modern required changes to the urban milk supply and dairying landscape. It involved the actions of public health officers and delivery truck drivers, farm families and food chemists, lawmakers and consumer activists, pediatricians and parents, bacterial cultures and dairy cows. To understand milk's twentieth-century transformation requires traversing the commodity chain, seeing the relationships between changes in consumer choices and farm practice. What emerges from such a history is not a simple story about milk, but a history of the evolution of consumer society, the development of governance over food and agriculture, the role of industrial technologies in organizing modern life, and the ways in which these processes engendered new ways of understanding nature.

Concern about food purity is not merely a historical phenomenon. Despite efforts by farm families, food processors, and state regulators to protect food safety, impurities continue to plague the nation's food supply—from spinach laced with *E. coli* to eggs contaminated with salmonella. Worried by disease outbreaks and increasingly distant from the farm, many Americans have, in recent years, taken renewed interest in learning more about the food system on which their lives depend. Their curiosity has turned books like Michael Pollan's *Omnivore's Dilemma* and Barbara Kingsolver's *Animal, Vegetable, Miracle* into best-selling tomes.[19]

Contemporary food writers such as Pollan and Kingsolver tend to characterize the food system as one of multiple paths: one of modern industrial agribusiness and another in which farming takes place "close to nature." The history of milk, though, reveals that the values and practices guiding industrial agribusiness and small-scale farming have not always been separate and distinct. Whether they fertilize fields with manure or synthetic fertilizers, farm families feel the forces of nature acutely when droughts or insect pests threatened to wipe out a hay crop. When raw milk devotees travel hundreds of miles seeking an unpasteurized product, their quest is just as embedded in the technological system and complex calculus of consumer demand that brings pasteurized milk to the nearby supermarket.

It is tempting to believe that nature can be controlled and equally alluring to be inspired to go back to nature. Milk's history reminds us that neither alternative is truly possible. Even at the moments when technology seems to guarantee new breakthroughs in managing and predicting processes of life on the farm, nature offers such challenges as storms, aborted cattle, and antibiotic-resistant bacteria. Similarly, even milk produced by pasture-grazed cattle, free of chemical inputs, carries residues from human activities. The pursuit of purity requires striking a balance between harnessing the raw materials of nature and allowing biological processes to thrive. To take milk's history seriously is to understand the compromises, complexity, and challenges involved in our dependence on other organisms for our very sustenance.

1

Reforming a Perilous Product

Milk in the Progressive Era

Strange though it may seem to us today, parents and physicians at the turn of the twentieth century looked at a glass of milk with far more fear and trepidation than we do today. When Progressive Era Americans encountered milk, they came in contact with a beverage bearing little resemblance to milk as we now know it. Milk contained clods of dirt and had a barny flavor that offended the dictums of good taste. Even seemingly clean milk harbored so many microscopic organisms that physicians described the food as a "living fluid."[1] Some bacteria gave milk a slimy consistency, increasing its viscosity so dramatically that it could be pulled into strings. Other bacteria colored it blue, green, or red. The most troubling bacteria of all were disease pathogens, and turn-of-the-century milk drinkers risked contracting a host of diseases, including diphtheria, septic sore throat, typhoid, bovine tuberculosis, and a variety of digestive distresses. During summertime heat waves, spoiled milk killed thousands of children, so many fatalities that some public health officials even viewed summertime infant deaths as a normal occurrence.[2] S. Josephine Baker, chief of the Bureau of Child Hygiene in New York City's Department of Public Health, would later recall that at the turn of the century, "telling a tenement mother to give her child so much milk a day was like telling her to give him a diluted germ-culture daily."[3]

Between 1900 and 1920, the alimentary pathway that once constituted a private link between mother and child became a matter of intense public concern. The hazards of milk would have been troubling enough by themselves, but they were compounded by social trends during the Progressive Era. As more mothers shifted from breast to bottle for infant feeding, more babies came in contact with the potentially harmful food. Urbanization magnified milk's perils, because as the food traveled greater distances from country to city, it stood a greater chance of becoming tainted. Poor residents in the nation's burgeoning industrial cities, who purchased milk in bulk and often lacked adequate ice to keep it cool, faced the greatest hazards. But milk-borne diseases struck wealthy residents as well as poor ones. Moreover, even though immigrants and the poor suffered disproportionately from public health ills, some reformers expressed greater concern about infant mortality among upper-class, native-born families. Such physicians worried that when preventable diseases such as typhoid or tuberculosis claimed the lives of rosy-cheeked infants of the "better class," it signaled the degeneration of American civilization. By ensuring that all of the "best" babies survived, they might shore up the authority and power of native-born Americans at a time of rapid immigration.[4]

If the problems posed by milk were rooted in the demographic shifts of the Progressive Era, the solutions proposed to purify milk followed the period's political trends. The early twentieth century was a time in which Americans came to understand consumer protection as a fundamental responsibility of the state.[5] Further, it was a time when women, extending their roles as food provisioners and caretakers of children, brought food purity and children's health to the public sphere.[6] Like most Progressives, women interested in improving the milk supply embraced a model of reform that melded grassroots activities with expertise. To ensure that the milk in their personal pantries was pure, they turned to experts and government officials to elucidate the wisest purchasing decisions. The shift from breast to bottle increased mothers' reliance on experts for infant care not simply by empowering physicians or infant food manufacturers, but also by enlarging the role of state inspectors as guardians of milk's purity.[7]

Two sets of experts were especially important to milk reform. First, public health departments took an increasingly active role in monitoring milk quality. By 1900, most urban health departments included at least one physician, engineer, epidemiologist, and chemist, as well as a cadre of inspectors. Public pressure, combined with new techniques to assess milk quality, made milk reform a central plank of municipal health activities. Milk was among the first substances to be analyzed in the lab with bacteriological methods, and milk safety played a central role in the health officials' efforts to reduce infant mortality.[8] During the first two decades of the twentieth century, public health reformers tested samples of milk to ensure its safety, licensed dairies, and enforced tuberculin-testing and pasteurization ordinances.

It is not surprising that public health officials sought to guard citizens from unsafe milk. What may be less obvious is that agricultural scientists also played an important role in improving the milk supply. The process of making milk safe required not just that the food itself be altered, but also that the dairy farms from which it came be transformed. Like public health, agricultural science had roots in the nineteenth century with the founding of land grant colleges and agricultural experiment stations. In the twentieth century, the Adams Act (1906) increased the federal appropriation for agricultural research and the Smith-Lever Act (1914) provided funds for researchers to share their insights with farm families through educational outreach. Agricultural scientists worked closely with rural elites, such as editors of farm papers, merchants, bankers, and implement dealers, to promote the adoption of new technologies to increase milk production and promote animal health.[9] The risks posed by milk were great, but health officials and agricultural experts expressed confidence that by applying scientific methods to dairying, passing regulatory acts that targeted milk adulteration, and educating Americans to take proper care of milk, they could radically improve milk quality and the efficiency of the farms from which it came. Their goal was nothing less than to transform both city and country and the white liquid that tied the two together. Ultimately, the campaign to make milk safe remade agricultural landscapes. The turn from human milk to cows' milk enlarged the scope of public health to reform the nature of the rural environment. Clean milk reformers, once most concerned with the households in which babies were raised, now scrutinized the qualities of the rural landscape and the practices of raising cattle. Through efforts to improve urban milk supplies, Progressive Era experts utilized state power to change infant feeding practices and human relationships to nature.

Environmental historians detailing changes to nature have more often traced changes to wild or urban landscapes than rural environments to understand the competing and contradictory ways that Americans approached nature in the Progressive Era. On one hand, conservationists employed engineering and social planning principles to use nature. Relying on scientific expertise, conservationists created forest reserves to protect watersheds and designed irrigation systems to distribute water.[10] Preservationists, meanwhile, sought to protect selected parts of nature from economic development and to set it aside as wilderness. By doing so, wilderness enthusiasts sought to restore to American men in particular the masculine vigor and independence they believed city life threatened.[11] The battle for wilderness was, in this traditional narrative, a fight between diametrically opposed groups: wilderness preservationists who valued nature for its aesthetics, and conservationists who valued nature for its usefulness.

But the history of milk and dairy farming in the Progressive Era demonstrates the ways that values of economic efficiency and natural aesthetics overlapped and intersected. On the farm and in the city, Americans describing milk and the landscapes that produced it celebrated nature even as they delighted in transforming it. They called cows' milk "natural" to link it to a vision of a peaceful and healthy countryside that contrasted with the corrupt and poisoned city and urged urbanites to drink pure "country" milk as an antidote to over-civilization. But public health and dairy experts also used the word "nature" more disparagingly to indicate lack of refinement. They feared the uncivilized tendencies of nature and thought it needed perfecting. Left to nature, cows would breed indiscriminately, eat weedy grasses, and produce only enough milk to feed young calves. Bacteria in milk would thrive and spread diseases to those who drank it. Seen through this lens, nature was merely raw material to be transformed into civilization. Though they credited nature for milk's purity, ironically, reformers altered the very nature of milk and the cows that produced it. The same people who described milk or cows as authentically natural also claimed that nature needed improvement. They declared their love for nature and then in the next breath demanded that it be changed.

By making milk safe to drink, as by establishing national parks or national forests, Progressive Era Americans sought to ameliorate the demands of modern, urbanized industrialism and create a place for nature in human life. Purifying milk, however, required reformers to draw upon the modern, urban, and industrial culture for which a glass of country milk was often seen as an antidote. Milk reformers turned to the new science of bacteriology to identify diseased cattle. They expanded the power of urban health departments to inspect farms and required milk distributors to purchase new equipment, such as pasteurizers. Never before had cows' bodies or the milk they secreted been so intertwined with human intentions and technologies. The pursuit of purity and the burgeoning consumer culture drove this delicate interweaving of technology and nature.

From Mothers' Breast to the Dairying Landscape

By 1900, most milk traveled a circuitous path from cow's udder to baby's stomach. Few urban Americans still kept their own cows, as they had in past centuries; rather, dairy farmers outside the city supplied them with milk and milk products.[12] Farmers located near the city might deliver it directly to consumers, but more commonly, farm families sold their product to a milk dealer who collected milk from a wide swath of rural territory and distributed it in the city. As Americans concentrated in cities and municipal borderlands urbanized, city residents came to depend on ever-more far-flung farms to slake their thirst. Many dairying families sent their milk to milk dealers aboard railcars, which often lacked sufficient ice to keep the milk cold. After reaching the city, milk could sit at railroad depots for hours before the dealer received it, giving bacteria ample opportunities to multiply. Then, the dealer pooled, bottled, and delivered milk to retail stores or to consumers' residences by horse cart. Milk sat in shop keepers' cans and on sidewalk stoops, sometimes covered and rarely iced. It is no wonder that Harvard professor Milton Rosenau commented in 1912 that "city milk, stale, dirty, and bacteria-laden, is therefore a very different article from the fresh country brand."[13]

If urbanization and industrialization allowed milk to travel longer distances between farm and market, ideas about the city and industry also spurred the decline of breastfeeding, which intensified the problem of tainted milk. By the turn of the century, fewer women breastfed infants than ever before, and those who did nurse weaned them more quickly, usually at or before three months. Still more mothers who breastfed supplemented their babies' diet with other foods.[14] Women shifted from breast to bottle in part because they worried that they had inadequate breast milk to feed their babies. Such anxieties derived from women's adjustment to urban life and their turn to more "industrial" models of motherhood. Some women turned away from breastfeeding because physicians claimed that urban life overstimulated women's nerves, undermining their capacity to create milk. In 1906, for instance, J. H. Mason Knox, who would later become the first president of the American Association for Study and Prevention of Infant Mortality, advised Johns Hopkins nurses that "the wear and tear of modern life, with its demands upon the mother's nervous strength and upon her time, and other factors less definitely recognized, have made it impossible for the human race to offer its progeny the sustenance intended by nature," that is, breast milk.[15] Other mothers sought to bring more regularity and order to infant feeding. Emulating factory engineers, they adopted strict feeding schedules—feeding babies only every four or five hours—in hopes of optimizing efficiency and teaching children self-control. But mothers' efforts ended the constant suckling that stimulated milk production and left them without sufficient milk to feed their children. Meanwhile, food manufacturers cast suspicion on the nutritional adequacy of breast milk by arguing that formulas provided babies with scientifically tested sustenance.[16]

The belief that civilized women were physically incapable of providing adequate milk characterized breastfeeding as a savage act and destigmatized bottle-feeding. By 1910, many citizens harbored a perception that as women became more civilized, their capacity to breastfeed diminished. As one observer noted:

> Like procreation, nursing is a purely animal function; and like the power to procreate, the ability to nurse offspring declines in the most marked degree among the leisured, cultured, and well-to-do classes. Among savages, as already noted, the condition is practically unknown.... But as we ascend the scale of civilization it is met with in ever increasing degree.[17]

Hence, despite efforts to convince mothers that breast milk was "best for baby," women began to believe that they could care best for infants by feeding them by bottle, not breast.[18]

Faced with rising infant mortality caused partly by children's increased exposure to bacteria-laden cows' milk, reformers could take up a variety of alternatives to improve milk quality. A first option that health reformers pursued was to discourage the use of cows' milk altogether. Even as breastfeeding became less widely practiced, many physicians tried to convince mothers that breast milk was superior to other infant foods, because it was more readily digestible and protected babies from infectious diseases. In working-class neighborhoods, nurses and public health workers promoted breastfeeding in prenatal home visits and at infant welfare stations. In 1909, physician Ira Wile, a consultant to the New York City milk depots, remarked that "the ultimate aim is to make the milk depots unnecessary institutions by teaching the mothers to nurse the children."[19] A physician who was part of the Boston milk committee even urged that wage-earning mothers be granted a pension so they could stay home and breastfeed their infants. Such physicians and public health workers believed breastfeeding would prevent the high rates of infant mortality, which spiked in the summer months when cows' milk spoiled.[20]

However, efforts to promote breastfeeding and discourage use of cows' milk were often stymied by the competing efforts of health officials to make cows' milk clean and safe to drink. For instance, when health reformers offered training sessions on bottle sanitation and pasteurization to girls charged with caring for younger siblings, their mothers became convinced that bottle-feeding had health officials' endorsement.[21] Although some pediatricians urged mothers to have babies suckle on the breast rather than nurse from the bottle, others found they could enlarge their scope of influence and enhance their professional authority by advising women on proper techniques to feed formula to babies.[22] Thus, despite a hard-fought campaign to promote breastfeeding, the use of cows' milk for infant feeding only became more entrenched in the first few decades of the twentieth century.

A second way health officers might have reduced the risks of milk contamination would have been to encourage the development of dairies within cities. In theory, city health departments could keep a closer watch on herds within their jurisdiction, and consumers would receive milk without the transportation delays that increased milk's bacterial content. This was not the strategy health officers pursued. By 1900, the presence of cows in American cities was steadily diminishing, due in large part to high property taxes and lack of pasturage that discouraged dairying in urban districts.[23]

Economic factors were not the only ones to discourage urban dairying. Among health officials, milk from urban stables suffered a poor reputation. In the mid-nineteenth century, breweries and distilleries in many American cities sold dairies the byproduct of their business—fermented brewers' grains—to use as cow feed. To some farmers, distillery slop, also known as "swill," seemed to be a boon; the feed came at low cost and stimulated milk production. But swill compromised cows' health and diminished the quality of the milk they produced. Fungus grew on cows' hooves. Their teeth fell out. Temperance reformers such as Robert Hartley began critiquing the practice of feeding fermented brewers' grains and "swill milk" in the 1840s.[24] Turn-of-the-century health officials and agricultural reformers picked up the fight against swill milk, making city-produced milk and swill milk synonymous.[25]

The tight association health officers drew between urban dairying and dishonest and unhealthy practices accelerated the expulsion of dairy farms from cities. In 1904, the United States Supreme Court upheld the right of the city of St. Louis to forbid cow stables within the city limits, arguing that "the keeping of cow stables and dairies is not only likely to be offensive to neighbors, but is too often made an excuse for the supply of impure milk from cows which are fed upon unhealthful food, such as refuse from distilleries."[26] Like reformers of the mid-nineteenth century, Progressive reformers associated the city with moral depravity and disorder. For those who considered cleanliness a virtue, the sordid state of urban neighborhoods marked them as spaces of vice.[27]

As urban milk reformers rejected urban dairying and failed to revive breast-feeding, they turned to "country milk" as the best option for infant feeding. So-called country milk appealed to early twentieth-century Americans because it came from the city's perceived antithesis. In contrast to urban environments, where tall buildings cast shadows and the masses of human life emitted a stench, the wide-open countryside seemed full of sunlight and fresh air. Thus, despite problems caused by unrefrigerated milk trains and unsanitary farming methods, the phrase "country milk" boosted sales.

The resonance of terms like "country milk" also had grounding in economic, political, and cultural developments of the Progressive Era. As mothers shifted from breast to bottle, women—once providers and producers of infants' milk themselves—began to encounter milk as consumers. Whereas mothers who breastfed knew the origin of milk intimately, those reliant on bottle-feeding did not. Instead, milk consumers relied upon expert feeding recommendations and neighbors' advice and evaluated advertisements as well as milk's cost, to decide whether milk was fit to feed their babies.[28] Descriptive categories like "country milk" addressed consumers' concerns about purity and the conditions under which food was produced.

For many milk drinkers, the phrase "country milk" appealed because the path from farm to icebox was anything but transparent. Unlike goods made and marketed by a single manufacturer, milk bottled under a dairy's label originated from hundreds of farms. Further, although packaged goods increasingly reached consumers under brand names, milk resisted branding. Working-class residents often bought milk from a bulk container without knowing which dairy company sold it. Other milk drinkers obtained milk via delivery and considered the availability of delivery service as much as the name embossed on the bottle. Instead of using brand names, consumers distinguished between milks by categorizing them: loose or bottled, pasteurized or raw, certified or not. By buying country milk, pasteurized milk, or certified milk consumers entrusted a place, technology, or process of monitoring to protect its healthfulness.[29]

Calling the food "country milk" also distracted consumers from a truth they increasingly understood: as the commodity chain to bring milk from cow's udder to baby's lips lengthened, the possibilities of the food's contamination increased. Health reformers noted the many avenues for potential contamination after milk left the farm and especially excoriated transporters, bottlers, and retailers for impure milk. Milk bottlers and distributors themselves sometimes brought such concerns to light. In its 1901 advertising campaign, Century Milk Company invited potential customers to come investigate its state-of-the-art bottle sanitizing equipment and intimated that few competitors properly sanitized their bottles.[30] Consumer groups, such as the Housewives' League, viewed the lengthening chain between farm and market with suspicion and sought to obtain products directly from farmers at a lower cost instead.[31] Even though so-called country milk traveled through many small town locales and urban places before reaching the consumer, the name "country" milk suggested it came fresh from the farm.

Images of rural life that circulated in the popular and agricultural press reinforced a common understanding of the country as a hardy healthful place that was free of urban vices and cast milk as the pure product of nature. In advertising images, cows grazed on verdant pastures or drank peacefully along meandering streams. Milkmaids toted pails of milk through the pasture, suggesting that milking took place en plein air and that women closely guarded the milk drawn from cows' udders. Such idyllic depictions hardly captured the reality of dairy farming, but by portraying cows' milk as close to nature and the milkmaid as its caretaker, milk companies instilled parents' trust in the safety of the food.[32] Images of cow's milk emphasized its natural and feminized qualities, perhaps to overcome misgivings about the decline of breastfeeding. In this way, commercial representations of the dairying landscape stood apart from advertising trends that identified the factory as the source of abundance.[33]

The sites in which Progressive Era Americans consumed milk furthered the association of pure milk with pastoral landscapes. When urban residents escaped the city during the summertime, milk was among the foods they sought out on such vacations. Catskills resorts advertised the fresh milk they served to entice urban New Yorkers to visit.[34] Similarly, some of the earliest pure milk reformers chose public parks as distribution sites for furnishing pasteurized milk to urban mothers and children. The aims of park creators and pure milk reformers appear disparate, but in fact both groups believed that their work brought the transformative powers of "nature" to the city. As park founders planted trees, milk reformers supplied urban residents with the purifying essence of rural nature.

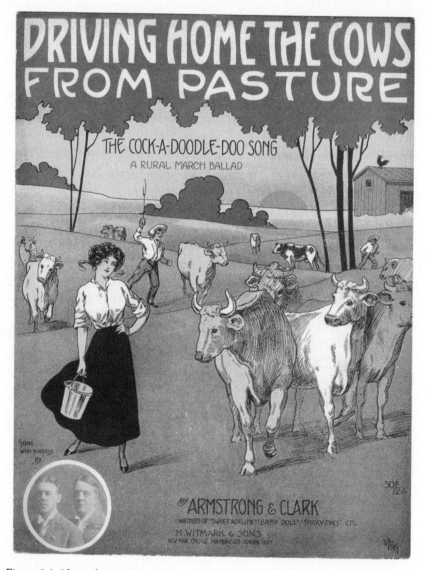

Figure 1.1 Ideas of pastoral nature helped assuage consumers' doubts about artificial food. The central role of women in these images made the transition from breast milk to cows' milk in infant feeding less dramatic. Sam de Vincent Collection of Illustrated Sheet Music, Archives Center, National Museum of American History, Smithsonian Institution.

Surrounded by pleasant trees and romantic hedges, city kids could sip nature's most nearly perfect food. As the park landscape soothed their souls, the fluid sustained their bodies.[35]

For turn-of-the-century urbanites, then, drinking milk was a way of seeking nature's cure. At a time when physicians considered exposure to pure air and sunlight to be the best cures for diseases, such as tuberculosis and hay fever, some patients afflicted with these illnesses sought out natural sites to improve their constitutions and added sleeping porches to their residences.[36] While not every consumptive could afford to escape the city, many could regularly ingest the products of the countryside.[37] Seeing milk as bottled sunshine and fresh air only amplified its reputation for healthfulness.

As they examined dairy farms, however, health officials quickly concluded that the designation "country milk" led consumers to envision milk's origins in a purer and more virtuous light than was the reality. Even if its origins were rural, country milk rarely derived from locales as idyllic as the countryside of the popular imagination. Just as many cows tromped through mucky ditches and stagnant ponds as grazed on verdant meadows near glistening streams. Many cows ate in poorly ventilated stables that they shared with chickens, hogs, and horses, not on the pasture. Unclean milk pails served as breeding grounds for bacteria.[38] The gap between dairy farm as idealized and realized motivated health reformers and agricultural experts to pursue a third strategy for improving the milk supply—transforming the physical nature of the dairy farm and altering farming techniques to guard the safety of milk from farm to consumer.

Purifying Milk: Implementing and Enforcing Reform

The health reformers who tackled problems with impure milk did so at a pivotal time in public health, in which the germ theory of disease provided sanitarians with a new tool to identify the food's problems. The environment in which milk was produced remained relevant to health reformers' assessment of milk's purity even as bacteriology and laboratory methods revolutionized public health work.[39] Dairy inspectors looked for cobwebs and sniffed for odors in the barnyard, as well as relying on laboratory tests to determine whether milk was free of disease bacteria.[40] Officials blended the tenets of germ theory with elements of sanitary practice and environmental control. This blend can be seen in each of the three methods for assessing milk safety between 1890 and 1930: dairy scorecards, bacterial counts of milk, and testing cows for bovine tuberculosis.

The first move toward dairy farm inspection began in the 1880s, as cities started to require farms to acquire a license to sell milk. To obtain a license, farm owners had to allow health officers to evaluate the working conditions of their farm and provide samples of their product. Health departments could revoke farmers' licenses or charge a fine if they believed a dairy was responsible for selling adulterated milk.[41] By the turn of the century, health inspectors sought not only to guard milk from adulteration with water or preservatives, but also to keep it free of dirt and disease. To this end, they introduced dairy scorecards to ensure fair and thorough farm investigations that would educate dairy farmers about how to produce cleaner milk. Scorecards assigned points to different elements of farm practice and equipment that contributed to milk's healthfulness. Most scorecards included provisions for the quality of the stable, the condition of the cows, the cleanliness of the cow yard, the equipment provided, and personnel's attentiveness to milking and keeping milk cool.[42]

Health officials' scorecards honed in on elements of the farm identified with the spread of disease in the city: poor ventilation, insufficient waste management, poor water supply, and presence of disease carriers. For instance, health officials believed a well-ventilated barn was of acute importance, particularly in northern climates where cows inhabited barns for much of the winter. Just as a resident of a crowded tenement district might be more likely to contract a disease, a cow in a crowded dairy barn could be more easily infected by other animals. Without proper ventilation, moisture accumulated and mold grew, compromising animal health and/or contaminating milk. Officials gave the highest score to barns that adopted the systems of ventilation that brought fresh air into barns and pushed animal fumes out.[43] Remembering how sewage had carried typhoid and cholera through nineteenth-century cities, health officials also took special interest in the manure pile.[44] They worried about flies flitting from a manure pile to a milk pail and spreading diarrheal diseases or typhoid fever.[45] They feared that manure that clung to the flanks of cows in an unclean barnyard would drop into milk pails. Farmers who regularly removed manure from the barnyard won inspectors' praise.

As some cities adopted scorecards to pinpoint the sources of milk's impurities, others utilized bacterial counts to evaluate milk quality. In 1900, New York City's Board of Health prohibited the sale of any milk whose contents exceeded 1,000,000 bacteria per cubic centimeter. Soon, Boston and Rochester officials set their own bacterial thresholds.[46] Health officers prized bacteriological examination for its efficiency. As Dr. Charles North explained, "One laboratory worker can test the milk of fifty dairy farms for bacteria while one dairy inspector is inspecting five dairy farms."[47] Proponents of bacterial testing believed that a plate count provided a more reliable gauge of milk quality than farm inspection, for a farm's rating by scorecard depended greatly on what inspectors witnessed during their visits. Unless inspectors happened to arrive at the farm during milking time, they had no definitive way of knowing whether dairy families washed and dried cows' udders before milking or cleaned milk utensils immediately after use.[48]

However, bacteriological testing also had faults. It took more than one laboratory test for inspectors to judge accurately a farm's milk. On a hot day, even a farm with relatively clean procedures might produce milk with a high bacterial count. More important, though bacteriological testing allowed inspectors to find out the aggregate number of bacteria in a sample, laboratory scientists were not yet able to determine which kinds of bacteria a sample harbored. Milk with an acceptably low bacterial count might nevertheless carry the germs of typhoid fever; milk with a high bacterial count could be free of disease-causing pathogens. Bacteriological analysis could approximate milk's risks but offered no guarantee of milk's safety.[49] Nor could bacteria counts confirm the precise source from which disease or taint in milk came.

Thus, laboratory examination rarely replaced farm inspection. Instead, many states and municipalities adopted a combination scale that included both bacteriological analysis and farm inspection. Bacterial counts served as a screening system, prompting future investigations to detect unclean conditions, diseased animals, or poor milk cooling practices. Dairy scorecards focused on the aesthetic elements and olfactory benefits of a well-tended farm, even when it proved to have little direct impact on bacterial contents of milk. Thus, as late as 1935, a milk inspector in Los Angeles initiated a program to improve the appearance of farms to passersby. Since many southern California dairies were located in close proximity to suburban neighborhoods where milk consumers lived, the visible and aesthetic appearance of farms played an important role in shaping milk drinkers' expectations of purity. The Dairy Roadside Appearance Program exhorted farm families to eliminate water holes and rubbish piles and encouraged them to plant flowers and paint farm buildings, in hopes of reassuring passersby of the quality of milk they consumed.[50] This continued aesthetic emphasis in a bacteriology-driven era demonstrates the lasting importance of physical characteristics of the farm to state officials' definition of milk purity.[51]

Whether evaluating milk by bacterial count or dairy scorecard, achieving milk purity demanded the cooperation and oversight of dairy farm families. Some scholars argue that Progressive Era urban health officials overpowered rural interests and favored the largest, most modern dairy farmers over middling ones. Other scholars stress that dairy farmers stubbornly refused to abide by newly established health laws, particularly regulations requiring cows to be tested for bovine tuberculosis.[52] Neither those that depict Progressive Era health reformers as powerful agents of an urban, capitalist agenda nor those who represent farm people as resisters of health reform fully capture the impact of milk regulations on rural people or environments. No matter how much power they wielded, urban health reformers could hardly ignore rural interests, because any hope of making substantive changes to dairy farm practice relied upon the cooperation of dairy producers. Further, because small producers dominated the Progressive Era dairy industry, milk reformers had to accommodate small and large farmers alike. Dairy farm families greeted health ordinances with a wide range of responses—from wholehearted acceptance to widespread resistance.

Farm people's reactions to health regulations depended on the economics of the dairy industry and on whether they accepted the methods used to gauge milk's purity as reliable and fair tools.

Many kinds of dairy farms supplied Progressive Era residents with a ready supply of milk. Some of the nation's most heavily capitalized dairies bordered American cities, where the ready market for milk, combined with the necessity of keeping milk cold en route, favored dairy specialization. Such farms boasted large herds of cattle bred especially for milk production and often relied on farm managers and hired labor to perform the tasks of milking, feeding, and tending cows. While milk for drinking was more likely to come from larger, specialized dairy farms than milk made into cheese or butter, a diverse lot of farm families contributed to the fluid milk supply of Progressive Era cities. Some milk producers kept just a few cows on general farms and used family labor to raise them. As urban ordinances expelled some small dairies from city limits, the expansion of rail transit enabled small and middling farms to send market milk to the city. For instance, in 1911, the average herd size of the 1,091 farms that supplied Washington, D.C., was just sixteen cows.[53] Even in areas where farmers specialized in dairying, such as the dairy districts of Pennsylvania, herd size rarely exceed fifteen cows as late as 1927. A more common herd size was five to ten animals.[54]

The persistence of small and general farms as dairy suppliers carried important consequences for farm people's receptivity to health regulations. On farms where the sale of dairy products generated one of many income streams, producers feared that implementing more labor-intensive milking regimens required by health codes would whittle away the meager profits that milk cows provided. Maintaining a profit margin on a mixed-income farm required economizing on the costs for dairy feed and housing. One University of Missouri study, for instance, noted that keeping cows in the basement of a horse barn would reduce building costs and help general farmers retain dairy profits.[55] But urban health officials condemned basement barns and those who housed hogs and horses alongside dairy cows, viewing the practice as unsanitary. Faced with contradictory advice about keeping the farm profitable and meeting sanitary requirements, dairy farm people weighed their roles as protectors of public health with their roles as businessmen and -women. Health officials sometimes blamed farmers' unwillingness to alter farm practice on ignorance or carelessness, but such inaction could just as well stem from devotion to Progressive Era business principles as from a disregard for health laws.

Even for large and well-capitalized dairy farms, the economic rewards for meeting high standards for milk purity were meager. Starting in the 1890s, physician Henry Coit began to promote the sale of premium-priced milk from dairies certified by committees of physicians called medical milk commissions.

Those dairy farmers with adequate means to adopt strict sanitary requirements and who produced milk with fewer than 10,000 bacteria per cubic centimeter could seek certification from the medical commission. Dairies that won approval could advertise their milk as "certified milk" and sell it at a higher price. However, few farms could afford the equipment and labor necessary to meet the certified milk standard. Dairymen who did sell certified milk found few consumers willing to pay premium prices. Nationally, certified milk was only available to consumers in the sixty-three cities where medical milk commissions operated and constituted less than 1 percent of the nation's milk supply.[56]

For the most part, then, the economics of dairying rewarded farmers for producing milk, regardless of its quality. Well into the twentieth century, the main measure of milk quality was its butterfat content, not its cleanliness. Since the price structure of dairying rarely rewarded efforts to produce clean milk, farm families concentrated more attention on increasing the amount of milk and butterfat their cows produced than improving milk's sanitary properties. Until the 1910s, when some cities adopted a system of milk grading, most dairy farmers received the same price for the milk regardless of how many bacteria it carried.[57]

When milk distributors *did* begin to reward dairies that produced milk with low bacterial counts or that attained favorable inspection scores with premium prices, farm families responded. In Geneva, New York, for instance, milk distributors began in 1909 to pay farmers whose barns and dairy buildings achieved a "good" score a half-cent more per quart of milk sold. Farms marked "excellent" earned a full cent more per quart. A half-cent per quart of milk was a modest price increase, but it often determined whether or not farms remained profitable. Economic rewards for producing higher-quality milk inspired farmers to pay greater attention to the health of their animals, to perform milking techniques cautiously, to remove manure more frequently from the barnyard, and to keep milk cool. By the end of 1909, a greater share of Geneva's dairies scored good or excellent than ever before.[58]

An important feature of the reforms instituted in Geneva was that health officials made the names of dairy farms and inspection scores public, enabling consumers to choose to purchase from farms deemed the most sanitary. Consumers took an even more active role in encouraging sanitary improvements in Portland, Maine. There, members of the Housewives' League published weekly statements of bacterial counts from each farm supplying the city. The League even rewarded farms that produced milk with the lowest bacterial counts with prizes, ranging from a heifer calf to a milk-bottling machine. Published dairy scores, like the white labels affixed to union goods by the National Consumer's League, empowered consumers to take part in achieving a more sanitary milk supply.[59]

The effect of paying for quality milk based on farm practice can also be seen in the records of Franklin Pope Wilson, who raised Guernsey cows near Purcellville, Virginia, and sold milk in Washington, D.C. With a herd numbering fifty milking cows, Wilson's farm, called Oak Knoll, exceeded the size of many farms supplying that city and was well equipped to meet increasingly stringent sanitary regulations. It boasted an ice house, a two-story dairy building with a cement floor, and a separate room for a cream separator, and it was equipped with both electricity and a water heater to help sanitize dairying utensils—features that set it apart from the average operation. Wilson himself resided in the city and relied on manager Charles Arnett to care for the farm.[60]

In regular correspondence, Wilson urged the farm manager to change practices to meet more exacting standards imposed by milk processors. In December 1921, one of the dairies to which Oak Knoll sold its milk—Wise Brothers' Chevy Chase Dairy—announced that due to a surplus of milk and consumer demand for a low-priced product it would cut the overall rates paid for milk. But Wise Brothers promised a bonus to any farms that achieved scores over 85 percent by the dairy scorecard. Wilson paid heed. In its September 1921 inspection, Oak Knoll Farm had scored an 82.3 percent. By improving the dairy's score just a few points, Wilson hoped to maintain high milk prices.[61]

Throughout the 1920s, Wilson carefully watched the dairy score and recommended minor changes that might improve the dairy's chances to obtain premium prices. In February 1922, Wilson advised Arnett, "It is going to be our policy now to get our dairy score up to 90 or above as soon as we can." He asked the manager to look for covered milking buckets, designed to prevent dirt or dust from falling into the milk pail, and reminded Arnett, "I see Dr. Drake has been giving us full score on the milk buckets, but he might stop it at any time if he sees we are not using them much." By June 1923, the dairy had garnered a score of 85.7 percent. To get the score any higher, Wilson believed, he would have to install a steam chest to sanitize dairy equipment. Equipment was but one factor measured by the scorecard. A visit in 1928, during which the milk inspector was "shaving us very close" prompted farm manager Arnett to give renewed attention to the more pedestrian elements of milk sanitation, such as caring for the manure pile. He reported to Wilson, "We are raking manure off the big lawn now, and it is quite tedious."[62] By tying farm conditions to milk prices, Wise Brothers' Dairy motivated farmers like Wilson to adopt new equipment and haul manure more frequently.

Maintaining a system that rewarded farmers who achieved high sanitary standards involved a complicated economic calculus involving farmers' profit margins, dairy distributors' supply of milk, and the public commitment to an improved milk supply. It was no coincidence that Wise Brothers' Dairy imposed a pricing structure to reward sanitary farm conditions during a period when

Washington, D.C., had a surplus of milk producers. The milk distributor knew that the risk of a milk shortage for enforcing more exacting health standards was slight, and farm families had added motivation to ensure they kept up a contract to sell milk at a good price. In sites where milk supplies were short, however, dairy distributors could not afford to be as exacting with sanitary provisions, for they needed all the milk they could get to maintain their business. Further, tying milk prices to farm improvements required accurate and frequent farm inspection. In 1911, when the city of Geneva's milk inspector resigned, a less experienced inspector began scoring the dairies supplying the city. Scores remained high, but conditions on the farms deteriorated. As the price obtained for milk ceased to reflect authentically the level of farm sanitation achieved, farmers' motivation to improve farm conditions eroded.[63]

Like farmers, milk inspectors adapted their approach to milk sanitation in light of economic and local conditions. Inspectors were thoroughly committed to improving the milk supply and protecting milk consumers from disease, but they also sought to create standards for milk purity that were practicable for farmers. Regular contact with dairy producers through on-farm inspection made some milk inspectors willing to compromise so that milk standards could be enforced. In some cities, such as Omaha, Nebraska, health officials treated dairies with only one or two cows less stringently than larger ones.[64] Recognizing that some recommendations were beyond the budget of ordinary farm families, dairy inspectors advised farmers on low-cost improvements that would make milk safe and reasonably sanitary, even if short of the scorecard ideal.[65] Small towns regularly lacked health departments or had such insufficient staffing that inspectors could not adequately monitor milk quality or test dairy workers for communicable diseases.[66] Strapped by budget limitations and lack of qualified personnel, the least labor-intensive inspection mode often prevailed. Even ambitious milk inspectors who sought to transform dairy practice and introduce bacteriological methods of testing milk reworked health ordinances to meet the needs of urban milk drinkers and rural producers alike.

Far from pushing an urban agenda, agricultural and health officials who acted to protect the milk supply, like companies that trumpeted "country milk," remained committed to the principle that rural life was more virtuous or preferable to life in the city. Rather than see themselves as abettors of an urban agenda, agricultural reformers saw their work of modernizing farms as a way to maintain the vibrancy and dynamism of rural life. As population increase in rural areas stagnated, agricultural reformers believed that making the practice of agriculture more scientific would prevent rural youth from seeking work in the city. By improving farm productivity and efficiency, reformers believed they could raise rural living standards and reinvigorate rural institutions. As historian Charles Rosenberg writes, "Experiment station scientists and administrators never considered the possibility that insofar as their work proved successful, it would mean helping to further enrich the rich, impoverishing many ordinary farmers and ultimately helping to force them from the land."[67] Rather than replace agrarian idealism with Progressive Era visions of scientific efficiency, reformers melded these visions, simultaneously encouraging radical changes in farming practice and seeking to keep young rural men and women on the farm.[68] Reformers sought change to ensure stability.

While economic conditions played a role in shaping how private dairies and public health officials executed milk safety laws, their influence was not simply to reward large farmers and drive small producers out of the dairy business. When small farms could implement new practices profitably, they did so readily. Dairy farmers' willingness and ability to improve practices to keep milk pure depended not only on whether they were economically able or progressive-minded enough to alter dairy practice, but also on whether the system used to regulate and purchase milk supported such efforts. Finding economically feasible milk sanitation solutions for farmer, milk processor, and inspector required give and take during implementation.

Translating Germ Theory to the Barnyard: Bovine Tuberculosis Eradication

Health officials who implemented licensing requirements and carried out score-card inspections assumed that the environment from which milk came accurately reflected its purity. However, when cows on seemingly well-sanitized farms were found to harbor disease pathogens, health inspectors and farm people alike were forced to rethink the relationship between milk purity and the environment. The discovery of diseases such as bovine tuberculosis on model farms did not supplant inspectors' consideration of nature in assessing milk, but it did shift their focus. They began to emphasize cows' bodies, not the countryside more generally, as the source of healthful milk. They analyzed the ecology of disease transmission and consulted with veterinarians to understand the relationship between animal and human health. In cases where bacteriological tests indicated disease, health and veterinary officials called for more radical changes, including the slaughter of infected animals.

To Progressive Era milk reformers, bovine tuberculosis seemed especially dangerous to milk safety. In 1908, a veterinary expert predicted that between 5 and 50 percent of dairy cows in the United States were infected with the disease.[69] Often bovine tuberculosis incubated in cows' bodies for weeks, months, or years before cows displayed clinical symptoms. Only as the disease reached an advanced stage would infected cows lose their appetite, produce a chronic cough, or develop swollen glands. Before these symptoms appeared, afflicted cows could pass bacilli through their milk or expel disease bacilli in manure and saliva.[70] The human health effects of bovine tuberculosis were serious. It caused intestinal infections and afflicted children disproportionately. In 1917, physicians predicted that 20 percent of tuberculosis cases in children under five were caused by bovine tuberculosis. In an era when one in ten Americans suffered from tuberculosis, health officials wanted to do anything in their power to prevent the disease's spread.[71]

Health and veterinary experts hotly debated the relationship between human and bovine tuberculosis and how best to control the disease.[72] They made two main proposals to curb bovine tuberculosis: eradicating cows that tested positive for bovine tuberculosis or pasteurizing milk. One way to stamp out bovine tuberculosis was to test all cows for the disease and kill all infected cattle. By 1890, bacteriologists developed a diagnostic tool to identify seemingly healthy cows that harbored tuberculosis bacilli. To determine if a cow was sick, testers injected a small quantity of tuberculin serum under the skin of the cow and monitored its temperature in two-hour increments. Cows whose temperature increased at least two degrees Fahrenheit within twenty-four hours were deemed infected.[73] The other method, pasteurization, required heating milk to 60°Celsius (140°F) for at least twenty minutes, which killed most disease bacteria communicable in milk, including tuberculosis bacilli. Milk dealers could pasteurize milk at the plant, or individuals could pasteurize milk at home. Pasteurization did nothing to eradicate tuberculosis among cattle, but it made milk safe to drink.

The history of tuberculin testing demonstrates how conversion to what historian Nancy Tomes calls "the gospel of germs" did not just change attitudes toward cleanliness. It also reconfigured the relationship between people and nature. The tuberculin test required farm families, veterinarians, and health officials to determine a cow's fitness on the basis of a bacteriological test, even when it contradicted conclusions drawn from clinical observation. Farmers were accustomed to assessing the health of animals by gauging the quality of their flesh, measuring the butterfat content of their milk, and observing their behavior. They believed cows that had good appetites, readily gave milk, and consistently bore calves to be healthy. When the tuberculin test contradicted observable indications, farmers were reluctant to kill healthy-looking prize animals.[74]

Even dairy farmers who agreed to have their cattle tested often doubted the veracity of the results. One Pennsylvania farmer was surprised when a cow's tuberculin test was negative, because the animal had a steady cough. Convinced of its infection, he killed the animal and found a piece of wire stuck in the cow's throat.[75] More often, bewildered farmers wrote agricultural officials when test results contradicted observed healthfulness. As one perplexed farmer explained,

"There is nothing about the animal to indicate the disease. She is perfectly healthy, in very good flesh.... Her milk is very high in butter fat. Would you kindly advise me whether you think this cow has tuberculosis or not?"[76] Another noted that he "didn't suspect anything...because their hair was all nice and sleek."[77] Insofar as farm people doubted the legitimacy of the tuberculin test, laws calling for mandatory tuberculin testing and the eradication of all reacting cattle appeared to be measures endorsing the slaughter of innocents.

Within this context, agricultural scientists convinced of the merits of tuberculin testing had to convert bacteriological principles into readily observable lessons. In 1905, bacteriologist Harry L. Russell convinced public audiences of the accuracy of the tuberculin test by staging public autopsies of cows that had tested positive for tuberculosis. As Russell dissected the bodies, observers saw the yellowish and gritty masses caused by the disease.[78] Postmortem demonstrations enabled farmers to apply a trusted technique of determining cows' health—observing its clinical symptoms—to tuberculosis bacilli that were externally invisible.[79]

As tuberculosis testing required farmers to rethink the ways they judged animal health, the procedure also called into question the assumptions that guided sanitarians who relied on inspection to verify milk's safety. Leaders of the milk sanitation movement, such as medical milk commissions, faced hard truths. In 1914, for instance, tests indicated that the cows on the farm of Stephen Francisco, the first farm to be approved to produce certified milk, suffered from bovine tuberculosis. Francisco's farm had long been lauded in the dairy press; its stables featured the best equipment and its laborers wore freshly washed white milking suits. It provided milk to hospitals and wealthy residents of New York, Newark, and Jersey City at a premium price.[80] When cows kept in a clean environment and certified to produce pure and healthy milk like Francisco's turned out to be receptacles of disease, it elicited concerns about practices used to ensure milk's safety.

In fact, farmers like Francisco were often more likely to house cows suffering from tuberculosis, because they frequently purchased cattle for breeding purposes to improve the average yield of their herd. Farmers who unwittingly purchased a healthy-looking infected cow suffered, as the sickened cows' excreta or saliva infected the rest of the herd.[81] Thus, although most often understood as a public health measure to protect urban milk drinkers, tuberculin testing was also a procedure that protected rural consumers who purchased livestock to increase their herds. Like a brand name on a package or a bottle of milk marked certified, papers that a cow was free of tuberculosis reassured a cattle breeder of an animal's quality, especially when purchasing animals from an unknown seller. As the broadening geography of consumerism inspired milk drinkers to seek out assurances of food safety, so too did it motivate livestock buyers and sellers to establish a way to ascertain that they were not purchasing damaged goods.

Historians often turn to the history of bovine tuberculosis eradication to underscore how vociferously farm people resented the broadening reach of urban power in the Progressive Era or to demonstrate the wide disconnect between scientific experts and lay people in the advent of germ theory.[82] Their emphasis is not misplaced. Many rural residents voiced frustration that urban health officials were deciding the methods farmers should use to raise their cattle. Even after state agencies granted indemnity payments for condemned livestock, farmers resented bearing the costs of testing and doubted that the price they received for tuberculin-tested milk would offset the expense of accrediting their herds as free of tuberculosis.[83] Farm people opposed to tuberculin testing mounted aggressive court challenges and even violently resisted local and federal officials charged with eradicating the disease.[84] Urban health officials who required tuberculosis testing found their efforts undermined by state legislators sympathetic to rural opponents who reversed local health ordinances.[85]

However, it is important not to characterize testing ordinances as purely urban measures or rural residents as wholly resistant. Rural people and agricultural interests played no small role in developing the legal and scientific infrastructure to eradicate tuberculosis. Agricultural leaders, particularly veterinarians and livestock breeders, had much at stake in efforts to eradicate the disease. Some farm people sought out testing. When Pennsylvania allowed farmers who suspected their cows to be tubercular to request testing, the board received three times as many applications for testing than they could perform.[86] Rather than pit urban against rural, the presence of bovine tuberculosis in dairy herds brought agricultural scientists into tighter partnerships with health officials than ever before.[87]

Further, tuberculin testing laws, like milk inspection laws more generally, were often reworked to meet local conditions and to grant rural people autonomy. Wisconsin policymakers, for instance, allowed individuals licensed by the state livestock sanitary board to administer tuberculin tests. Giving farmers who passed the licensing exam the authority to test cattle reduced farmers' inspection costs. More important, it provided farm people the opportunity to rely on a trusted, local community member to perform the test and thereby minimized concerns about expert outsiders interfering in rural affairs. By 1911, between three and four hundred nonveterinarian testers had passed the Livestock Sanitary Board's exams to become tuberculin testers.[88] That so many farm people sought out roles as tuberculin testers demonstrates that, despite resistance, many rural people were convinced of the necessity and accuracy of the tests. Nevertheless, progress in eliminating tubercular cattle through testing proceeded slowly, because after a positive test and destruction of an animal, herds had to be tested annually to ensure the disease had not infected another cow. Often, it took years for the disease to be eliminated from a single herd.[89]

Due to the slow pace of eradication on the farm, by the 1910s, more and more municipalities passed ordinances requiring pasteurization instead of relying on the tuberculin test to protect the milk supply from bovine tuberculosis. Unlike farm inspection and tuberculin testing, pasteurization directed the responsibility and cost for ensuring milk safety toward urban milk dealers and consumers and away from farm families. For pasteurization's advocates, the process through which milk was treated, rather than the origin of the product, was of key importance. No matter how many herds of cattle harbored disease, pasteurization could be applied to all milk sold in the city, thereby protecting all city residents from disease, not just discerning consumers who insisted their milk came from tuberculin-tested herds or who could pay premium prices.

Public health officials' turn to pasteurization began in the nineteenth century as a response to high rates of infant mortality among the urban poor. Nathan Straus, a philanthropist who believed that pasteurization efforts in Europe saved the lives of many infants, was a tireless pasteurization advocate. He set up milk depots to distribute pasteurized milk to New York City's children between 1893 and 1910.[90] Physicians dedicated to improving the lives of the urban infant poor, such as pediatrics pioneers Abraham Jacobi and J. H. Mason Knox, lauded and supported Straus's efforts.[91] Physicians and public health workers found that babies who drank pasteurized milk suffered less diarrhea than those who drank raw milk, especially during the summer.[92] For those troubled by high rates of infant mortality, pasteurization presented a means to reduce infant deaths across class lines.

If pasteurization appealed to physicians because it reduced infant mortality, health officials and agricultural experts believed that pasteurization was practicable and economical. For milk regulators, the time and expertise required to conduct farm inspections and the additional challenge of overcoming skeptics of tuberculin testing made cleaning up the milk supply difficult and expensive. Milk inspectors argued that improving dairy farms would drive up the price of milk because it required farmers to have adequate equipment and skilled labor. Few milk drinkers, they cautioned, paid to obtain certified milk. But pasteurization would cost "a meager 1/8 cent per quart."[93] To these proponents, pasteurization would diminish the cost of milk inspection and eliminate the potential headache of tuberculin testing, all the while keeping milk safe.

The medical community's support for pasteurization was by no means unanimous. Physicians were troubled by the effects of pasteurized milk on its nutritional quality, for heating milk killed not only pathogenic bacteria but also the lactic acid bacteria believed to be beneficial to digestion. As late as 1912, over half of pediatricians surveyed by the American Pediatric Society believed "babies did not thrive well on pasteurized milk and that such milk could lead to infant digestive disorders."[94] A few physicians argued that pasteurized milk made children prone to rickets and scurvy.[95] Another concern was that milk companies conducting pasteurization would not perform the process properly. By 1908, commercial milk plants in Cincinnati, New York, Philadelphia, Milwaukee, St. Louis, and Chicago had begun pasteurizing milk. But these milk plants did not pasteurize the milk long enough or at high-enough temperatures to kill disease-causing bacteria. Some companies pasteurized milk multiple times or used dirty pasteurizing equipment. By pasteurizing milk, they sought not to eliminate disease but to improve milk's keeping quality and to reduce the amount they spent for ice to keep it cool. Such examples did not give health officials confidence in the ability of milk companies to protect the milk supply.[96]

The most basic and important objection voiced by opponents of pasteurization was that truly healthy milk would not need to be pasteurized, because it would come from sanitary dairies and healthy cows. Physicians worried that pasteurization would discourage efforts to improve milk's cleanliness and sanitation on the farm, because it would make "dirty" milk passable for consumers. Pasteurization, they emphatically repeated, could make milk safe, but it could not make dirty milk clean. According to certified-milk advocate Henry Coit, widespread adoption of pasteurization would "postpone the time when a uniformly pure milk supply would be available to the masses."[97] Through this lens, pasteurized milk was unwholesome, not an authentically healthy food. In 1907, when the Los Angeles Health Board rejected a pasteurization ordinance, they called pasteurized milk a "doctored fluid."[98]

As experts debated its merits, ordinary citizens remained doubtful about pasteurized milk because the heating process of pasteurization altered one of the elements they used to evaluate milk's quality: the cream line. Before milk plants homogenized milk and low-fat diets became vogue, a deep line separating the cream from the rest of the milk marked it as healthful and rich. Some dairies sold their milk in "cream top" bottles, with indentations in their necks to exaggerate the cream line.[99] Pasteurization minimized the depth of the cream line. Therefore, as cities passed pasteurization ordinances, how to maintain the cream line became a key subject in the dairying press. Milk machinery companies touted pasteurizers that preserved the cream line more effectively than other brands.[100] New York City's Board of Health even revised its pasteurization rules so that dairies could perform pasteurization at a slightly lower temperature and maintain a well-defined cream line.[101] Just as inspectors responded and adapted health ordinances to take into consideration producers' concerns, consumers' desires for a clearly defined cream line indicate their influence on establishing standards for milk quality.

Ultimately, Americans did not end up choosing between the practices of tuberculin testing or pasteurization for milk purification, but combined the approaches. On the local level, municipalities began to require that all milk, except for certified milk from tuberculin-tested herds, be pasteurized. Chicago was the first to pass such an ordinance in 1908. By 1916, pasteurized milk constituted 90 percent of the milk sold in New York, 85 percent of that sold in Philadelphia, and 80 percent of Chicago's and Boston's supply. Though progressing more slowly in small cities and rural areas, by the mid-1930s, 98 percent of the urban milk supply was pasteurized.[102] As pasteurization ordinances took effect, agricultural officials worked to systematically eradicate the incidence of bovine tuberculosis in livestock herds, stamping out the disease by killing reacting cattle and providing compensation for dairy farmers. Blending pasteurization and tuberculosis testing, agricultural and health officials eventually eradicated bovine tuberculosis and improved milk's safety.

Conclusion

Transformed by pasteurization and farm inspection, by 1920, milk became a revered staple food for infants and children.[103] World War I publicized milk's nutritive value to consumers, as Americans exported condensed milk to their allies in Europe and as nutritionists recommended that families eat dairy foods instead of wheat or meat. The National Dairy Council, founded in 1915, brought further acclaim to milk and its products. In 1918, pure foods advocate Harvey Wiley argued that access to milk would define Americans' future, writing, "Just in proportion as we can supply milk to the children shall we have healthy men and women in the next generation."[104] Instead of being viewed as deadly, even dangerous, cows' milk now seemed to guarantee good health.

Such a positive view of milk's healthfulness would have been impossible at the turn of the twentieth century. Turn-of-the-century health officials criticized milk for spoiling, stinking, and containing dirt. Rather than being known for spreading health, milk had been notorious for carrying bovine tuberculosis, typhoid, and scarlet fever to those who drank it. But by the 1920s, milk became popularly known by a new title: nature's perfect food.[105] Scientists and nutritionists contended that milk supplied its drinkers a more balanced mix of nutrients than any other food. That Americans began to characterize milk as a perfect food was remarkable, because milk had once very nearly been defined by its imperfections.

Milk's new image was not simply the work of dairy publicists. The food became safe to drink because agricultural and public health reformers urged farm families to change the practice of dairy farming—rebuilding barns, hauling manure, and testing cows to bring nature under control. Milk codes, campaigns to eradicate animal diseases, and pasteurization ordinances reconfigured the nature of dairy landscapes and milk itself.[106] Only once milk inspectors cleaned up dairy farms and pasteurization rid milk of disease, would they call milk not just a healthful food, but a perfect one.

By stressing milk's perfection, the phrase "nature's perfect food" broadcast the recent and radical changes to milk's quality. Yet by implying that milk was good because of its nature, the phrase also raised questions about the necessity for human intervention in natural processes. If milk was perfect by nature, would human efforts to improve nature undermine its perfection? As more Americans came to call milk "nature's perfect food" and transformations to milk intensified, the ambivalence the phrase implied about the appropriate role for humans in nature would only deepen.

Balancing the Goods of Nature

Butter in the Interwar Period

In the winter of 1911, a group of New York women launched a protest movement. The issue that motivated them was not suffrage, prohibition, or child labor. Rather, they were concerned about access to and quality of a household essential: butter. When commission houses hiked prices of the savory staple, the women refused to purchase even a pat. Within days, butter marketers dropped their prices. In the wake of the successful boycott, thousands of women nationwide joined the group who called themselves the Housewives' League.[1]

The price of butter may appear a peculiar instigator of a protest. But Americans in the 1910s and 1920s demanded butter, and lots of it. Whether they spread it on toast, used it to fry pancakes, or baked it in pies, between 1910 and 1940, Americans regularly ate more than eighteen pounds of butter per capita, well over four times the recent consumption rates.[2] In 1918, one magazine recommended that a household consume as many pounds of butter per week as there were members of the family.[3] For women who wanted to provision their tables with the best food and to rein in the grocery bill at a time of escalating prices, butter was of weighty importance.

A staple for American eaters, butter was also central to the farm economy. Many more rural families churned butter on the farm or supplied cream to butter factories than sold milk for drinking, because butter factories did not require farmers to meet as exacting sanitary requirements or as quick of a transit to market. By the 1930s, more milk was channeled into butter than to any other dairy food.[4] Hence, butter prices inspired farm people, as well as consumers, to organize.[5] High butter prices, the target of consumer cost-of-living protests in 1911, were the goal of farm families ten years later.

At first glance, the politics of butter pricing seem to confirm historians' assessment of the interwar period as a time of tense relations between country and city. During the 1920s, immigration, prohibition, and religion pitted cosmopolitan urbanites against rural fundamentalists. Improved roads, telephones, and new media—like advertising and radio—brought urban values and cultural forms to rural audiences, challenging local institutions and sensibilities.[6] But even as rural people voiced objections to urban mores, they sought out ways to reconcile urban industrialism with rural life. By 1920, more Americans were urban than rural, and rural people, especially youth, shunned the country and sought opportunities elsewhere. The nation's economy, once grounded in agriculture and raw materials, shifted to become capital-intensive and consumer-centered. Fearful of population drain, farm experts urged producers to change the nature of the farm to fit the demands of the standardized market. By the 1930s, farm producers increasingly believed that they needed to accommodate farming practices to meet the dictates of the consumer-driven age.

That even humble butter could be part of this push toward mass-marketing and standardization might have been difficult for early twentieth-century Americans to fathom. Before the 1920s, butter was manufactured largely on farms or by local creameries. The weight, color, quality, and volume of butter sent to market varied widely. Seasonal fluctuations in cows' diets, the ways individual farmers balanced their farm work, and the whims of the buttermaker meant that the body and flavor of butter differed in nearly every tub. Wholesale commission agents graded butter in Philadelphia and New York before selling it to retailers, but only a few buttermakers paid keen attention to the standards on which butter was judged. Many creameries even lacked scales to weigh their product before sending it to market.[7] Butter marketing was personal and localized. Farm kids delivered butter churned by their mothers to town residents on their way to school, and women peddled it directly to consumers at curbside farmers' markets.[8] For most of the farm families supplying the cream trade, milking cows was a seasonal sideline, not a main income stream. Cream sellers raised beef cattle, hogs, horses, and chickens. Many also sold fruits, potatoes, sugar beets, hops, or tobacco. Often, cream-selling farmers valued dairy cows as much for the manure, calves, and skim milk they produced as for their cream. Whether selling it from the farm or alongside other farm products, cream and butter sales offered farm families a measure of economic security in a volatile agricultural economy dominated by crops like cotton, tobacco, corn, and wheat.

During the interwar period, farm people and buttermakers changed the ways they made and marketed butter. By 1920, buttermaking largely shifted from farms to factories, called creameries. The largest of these factories, centralizer creameries, operated at the intersection of railroad lines.[9] Smaller, crossroads creameries thrived in regions of the amply populated upper Midwest where dairy farming thrived but farmers were too distant from cities to supply urban fluid milk markets.[10] By sending cream to the factory to be churned, farm people could lighten the labor of dairy work and benefit from the expertise of a skilled buttermaker.

The rise of the chain store also altered the butter trade. By 1923, nearly a quarter of the nation's food retailers were chain stores.[11] Driven by centralized management directives, chain stores purchased goods directly from manufacturers and rewarded those who could deliver a large volume of high-quality butter. Chain store managers wanted consumers to be able to enter a store in San Francisco, Spartanburg, or Schenectady and obtain the same kind of product. Centralizer creameries, because of the large volume of butter they produced, were well positioned to take advantage of the new retail environment. Local creameries could compete with centralizers if they adhered to a new standard of purity: one that promised a better flavor, longer-keeping quality, uniform color, and homogeneous texture. Chain stores also encouraged butter purchasers to use new ways to judge butter quality. Once attentive to the tub with the most appealing shade of yellow or the endorsement of the grocer, consumers came to choose butter based on labels carrying the imprint of state inspection and nutritional claims.

The most profound shift in the butter trade was resetting the priorities of the farm people who supplied it. Dairying experts asked small farmers to stop viewing their cows largely as generators of products to be used on the farm—like manure to fertilize soil or calves to increase the herd—and instead to value cows for the cream they produced. Local creameries expected farmers to be more fastidious in caring for cream and feeding animals to maximize production. Increasingly, success in farming required both careful understanding of the potential of one's farm, as well as market savvy about the national and regional demands for farm products. Rather than view interactions with regional and national markets as undermining autonomy, farm people saw the vitality of the local farm economy to be dependent on these interchanges.

From Farm to Factory: Mass Production via Small-Scale Producers

Contemporary rhetoric neatly divides local, small, subsistence farmers from nationalized, industrial ones. But a closer look at the historical record of buttermaking in the interwar period reveals ways that local farming and industrialized mass production often overlapped and coexisted. The experiences of Elihu and Lena Gifford, who farmed near Camden, in Oneida County, New York, were in many ways typical of those engaged in the cream trade. Throughout the 1910s and 1920s, the Giffords kept a herd of twenty to forty Jersey cows, a breed that yielded milk with high fat content and thus was perfect for a farm selling cream. Compared to other cream-producing farms, the Giffords' herd was large, but still cream only constituted a fraction of the farm income. Most winters, the Gifford family produced a hundred pounds of sausage from hogs they killed on the farm, slaughtered beef cattle for their own use and to sell to neighbors, and dressed scores of chickens for local grocers. The Gifford family also raised other livestock, not least of which were horses that aided in much of the farm work. In spring and summer, they delivered sweet corn to a nearby cannery, kept a garden, and raised oats, hay, corn, and beets for livestock fodder. Offering a variety of products for market helped balance the volatility of agriculture and steadied the income stream on the farm.[12]

Local food enthusiasts might be surprised by how many national ties kept farms like the Giffords' afloat. When Elihu and Lena Gifford ventured to the local store with farm-raised sausage, hams, eggs, and beef, they returned to the farm with pineapples, bananas, oranges, and lemons. The Giffords fed livestock farm-raised alfalfa hay, oats, corn silage, and beets, but also supplied them with oil meal and cottonseed meal shipped from St. Louis and points farther south. As Mr. Gifford sold cream and butter to nearby restaurants and local residents, he also traveled to New York City, Utica, and Syracuse to establish cream contracts with greater geographic reach. Supplementing local feed rations with cottonseed meal boosted production while keeping feed costs low. Securing a more lucrative cream contract in New York City guaranteed a higher return for one's labor than one might receive locally. The ledgers of farms like the Giffords' illustrate the careful balance of markets driven by neighbors and by national trends, and of combined uses of local resources and of inputs purchased from afar.[13]

The rise of centralizer creameries provides one of the most vivid examples of the coexistence of small-scale farming with the tenets of the industrial economy in the 1910s and 1920s. A new machine—the hand-cranked cream separator— fueled the shift in butter production from farm to factory. Before the twentieth century, most buttermakers relied on gravity to carry out the first step in making butter, which is separating whole milk into cream and skim milk. They poured milk into shallow pans, placed the pans in cool water, and waited for cream to rise. Then, they used a round disk to remove the cream layer from the skim milk.[14] In the 1870s, Danish and Swedish inventors devised implements to separate whole milk into cream and skim milk more efficiently. As whole milk whirled in a bowl, centrifugal force separated skim milk from the lighter, insoluble cream. The spinning motion pushed the heaviest parts of the milk to the edges of the bowl, while the lightest cream remained in the center. Once separated, skim milk and cream flowed from separate outlets. Butter factory operators adopted centrifugal separators because they could operate at any temperature and could control the richness of the cream they yielded.[15]

In the 1890s, a new kind of centrifugal cream separator came on the market: a hand-cranked centrifugal separator for farm use. Soon, cream separator companies, such as De Laval, Sharples, and Burrell, aggressively marketed them to rural audiences. The separator manufacturers peddled their wares at county, state, and world's fairs, canvassed the countryside, and advertised in all the farm magazines.[16] Dairying families responded quickly. When the De Laval separator salesmen visited the Gifford farm in 1910, for instance, they found the family already had a separator.[17] By 1930, creamery expert Otto Hunziker explains, "nearly every family in the dairy belt had a farm separator."[18]

The adoption of hand-cranked cream separators in the 1910s and 1920s made it easier than ever for farm families to supply butter factories. Historically, farms that engaged in the butter trade were more geographically distant from urban markets than those that specialized in the fluid milk trade because fluid milk producers needed milk to reach the city before perishing. Even creameries were reluctant to operate in scantly populated areas, for the potential volume of raw material was too small to justify the operating costs. The cost of hauling whole milk to the factory exceeded potential returns from cream sales. The farm separator changed the economic calculus. It enabled farmers with just a few cows to begin marketing their surplus butterfat. With its adoption, factory butter production extended into the most rural regions of the nation, including the Great Plains and the South. Large creameries took advantage of these small cream sellers who sent separated cream by rail to the factory.[19] By collecting hand-separated cream from far-flung places, centralizer creameries incorporated small-scale farmers into the national economy. Specialized modern processing supported small-scale mixed farming.

Despite the general trend toward factory production, in some rural districts, home-produced butter still commanded an important sector of the rural market. One 1915 study, for instance, found that nearly 90 percent of farmers in northern Minnesota made their own butter, including those who also sold cream to local butter factories. Locally made butter was consumed in the home but was also traded to storekeepers for credit. Most grocers and general store owners acquired more butter in these small transactions than could be used locally. Storekeepers sent surplus butter via merchants to centralizer creameries, where it would be remade and sold to urban consumers as "renovated" butter. To renovate, centralizers melted the products of hundreds of small country stores, combining butters of varying quality and age. After removing the whey and other impurities from the butter oil that remained, manufacturers blended the oil with skim milk or buttermilk so that the final product would have a similar appearance, consistency, and flavor to genuine butter.[20] Renovating butter allowed a poor-quality product—homemade country butter—to find a market, relieving the pressure on country storekeepers who accepted such butter in trade. In transactions like this one, the line between local and industrial, homemade and corporate became especially difficult to draw.

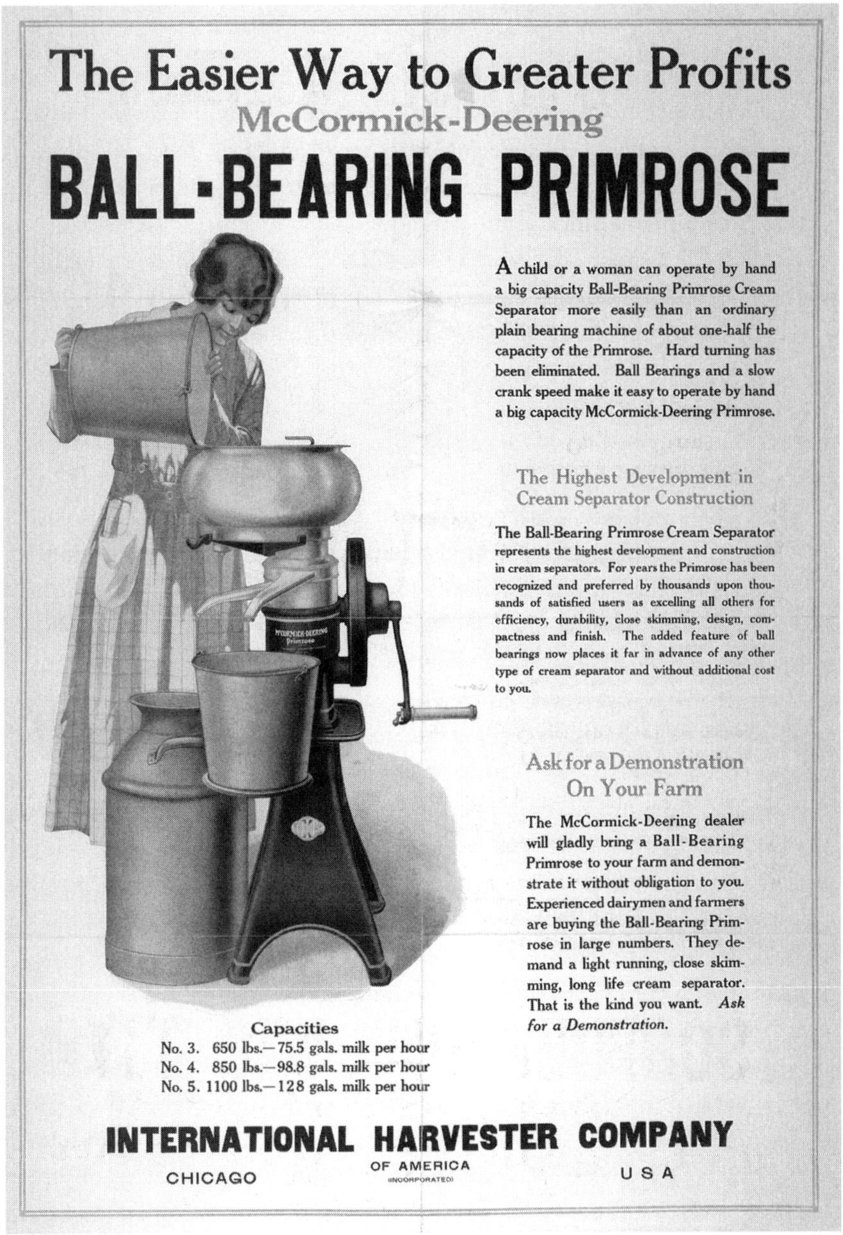

Figure 2.1 The cream separator enabled farm people to deliver cream to market and use skim milk on the farm for animal feeding. Keeping the separator clean was essential to the production of wholesome and sanitary butter but was a toilsome task. Wisconsin Historical Society, WHi-9964.

Cream Purity: An Environmental, Technological, and Social Problem

The farm separator helped farm families enter the butter trade, but it provided no guarantee that the cream they sent to be churned into butter would be wholesome and high-quality. Despite manufacturers' promises, mechanically separated cream often reached the factory dirty and spoiled. Interwar Americans seeking to improve cream attributed its poor quality to both the factories making butter and the farmers who supplied them. Some, foreshadowing present-day critiques of factory farming, blamed poor cream on the large size of centralizer creameries and the anonymity they afforded their farmer suppliers. To these critics, the scale of modern processing imperiled food quality. Others saw the seasonal and unspecialized nature of the farms delivering cream as the primary obstacle to purity and called for greater focus on dairy work and professionalization of buttermaking.

The factors to which observers traced cream's impurities and the solutions they suggested carried political implications. Efforts to purify cream took place at a time when many rural Americans sought to counter the growing power of agricultural processors by forming their own organizations to store or process agricultural goods. Many local creameries were cooperatives, owned jointly by the farmers who supplied them (called *patrons*).[21] Promoters of agricultural cooperatives believed that by selling the product of their labor in common, farm people could revive the economic vitality of the countryside and maintain a measure of autonomy. When cooperative proponents linked impure cream with the centralizer system, they cast aspersions on their competitor and critiqued the broader economic system of which it was a part. Similarly, when big butter factories derided "country butter," they characterized rural creameries as ill-run, backward-looking facilities.

Although local creameries and centralizers' rhetoric pitted them against one another, all butter factories faced many of the same kinds of challenges in obtaining good cream. It mattered little whether a centralizer or local cooperative churned the cream into butter if a heat wave melted the ice used to keep cream cool, or farm families skimped on the time devoted to sanitizing dairy equipment. Since many farmers sometimes sold cream to centralizers and sometimes to cooperatives, depending on the price they could obtain, causes of impure cream rooted on the farm affected centralizers and cooperatives alike. Poor cream resulted from a mix of natural, technological, and social variables. Some natural factors, like bacterial spoilage, could imperil cream, but other environmental variables, such as cool spring waters, could be harnessed to maintain its purity. Similarly, modern scientific and technological interventions could overcome the natural variability of milk but could also accelerate problems of spoilage by intensifying competition between creameries. To get pure butter, buttermakers had to balance natural processes and technological interventions.

One source of unsanitary cream was unwashed or poorly sanitized cream separators. The same intricate separator parts that channeled cream to the center of the bowl collected slimy milk residue. Without careful sanitation, vestiges of milk, and the germs it carried, remained. The task of separating milk, then, was not complete until farm people ran lukewarm water through the machine, disassembled the separator, washed each individual part, and sanitized them by scalding with hot water. Tiresome enough by itself, cleaning the separator also often required pumping and heating water. This task was almost always women's work. Farm women remembered years later how much they loathed cleaning the separator. Isabel Baumann, who grew up on a farm near Stoughton, Wisconsin, recalled, "My one job that I hated so much was to wash the separator, especially in the summertime.... I vowed that one day I'd never have to wash the separator and by gosh, I get married, and move up here there's the separator." Clara Scott, from Grant County, Wisconsin, also explained, "separators...were a terrible thing to wash. All them discs and everything."[22] If even a few women shirked the meticulous sanitation process, they imperiled the safety of the cream sent to market.

The site where cream separation took place could also affect the purity of the product. Many who marketed cream practiced mixed farming, and such farmers valued the skim milk coming from the separator as much as the cream. Skim milk could be fed to chickens, hogs, and calves, and enlarged the profitability of the farm's livestock, because the protein-rich skim milk could replace some of the high-cost protein-rich grains or supplements purchased for animal feed. Some farmers were so concerned with the quality of skim milk for animal feeding that they separated milk in the hog barn. This practice made feeding hogs convenient but offended milk inspectors' notions of cleanliness. In 1913 and 1914, milk inspectors in Wisconsin issued citations to farmers who kept their milk separators and cream in unclean parts of the barn, for the cream arrived at the creamery warm, full of particles of dirt, hay, and manure, and tainted with a barny flavor.[23]

Cream purity also suffered because health officials prioritized farm inspection of sites producing milk for drinking over those supplying cream to the butter trade. Food-borne illnesses from butter were less frequent than those traced to spoiled or contaminated milk. Furthermore, public health experts could identify and monitor suppliers of fluid milk more easily than cream producers. The sheer numbers of farmers that sent surplus cream to creameries, combined with the geographic isolation of these farms, made regulation difficult. Some farmers sold cream for butter only seasonally or sporadically, presenting an even greater challenge to health inspectors. Because dairying was merely a sideline operation on many cream-producing farms, such farmers were reluctant to invest in improved facilities for conducting dairy work. Most cream-producing farms were only visited by inspectors once a year, if at all.[24] In combination, lax regulation, mixed farming, and lazy separator sanitation could imperil the cream supply of cooperative and centralizer creameries alike.

Cream destined for centralizer creameries did travel longer distances and with less frequency than cream carried to neighborhood cooperative creameries by wagon. Since many of the farm families supplying centralizer creameries had just a cow or two, it sometimes took them a week before they gathered enough cream to merit a trip to the cream station. As the cream sat waiting to be delivered, it lacked proper cooling and sanitation, especially in the summer months. Failing to cool cream completely before adding it to the can of already-collected cream could result in a musty, smothered flavor. Uncovered cream cans also attracted flies. A 1935 FDA investigation found maggots in eight of the fifty cans of cream delivered to one southern creamery.[25] Transportation delays degraded cream quality even further. By the time cream arrived at centralizer creameries by rail, it had traveled long distances. Sometimes cream sat unrefrigerated at the railroad station before being loaded and delivered to the butter factory. Aaron Ihde, who worked at one of the largest centralizer creamery companies from 1931 to 1938, described the cream that arrived at Blue Valley Creamery in Chicago:

> Cream that was very sour sometimes had developed off flavors, sometimes had even accumulated insects and rodents. In those cases Blue Valley dumped the stuff and sent the farmer a note and usually lost them as a customer.... But every so often the rat didn't appear until it was dumped into a big vat.[26]

Blue Valley Creamery was not an isolated case. Floyd Lucia, who worked as a buttermaker for the Beatrice Creamery Company in Indiana during the early 1920s, noted that cream sometimes arrived at its factories nearly rancid.[27]

Critics of the large centralizer creameries blamed the long path from farm to market for discouraging rural people from being careful with their cream. In the centralizer system, the staff at the local cream station, not trained buttermakers, had the most contact with farmers. Drop-off stations for centralizer creameries were located at railroad stations, garages, gas stations, blacksmith shops, and private houses. The mechanics or railroad operators who staffed the stations were rarely trained in dairy sanitation. Cream station operators sought to increase the volume of cream sent, not to ensure its quality.[28]

At local creameries, buttermakers had regular contact with those who delivered milk, could advise them on problems with their cream, and warn them when cream did not meet expectations.[29] And yet even local buttermakers who had been trained in the principles of buttermaking and who enjoyed direct contact with farm patrons could lapse in their insistence on high-quality cream. While local buttermakers could instill sanitary practices and enforce a high purity standard, patrons exerted influence on buttermakers' actions, sometimes discouraging the buttermaker from doing so.[30] As G. L. McKay and C. Larsen noted in their 1922 textbook *Principles and Practice of Butter-making*:

> The question as to where the line should be drawn between the good, medium, and very bad milk or cream, must depend upon the judgment of the receiver, and in a great measure upon the *local conditions*.[31]

Some local patrons resisted buttermakers' requests for pure cream. M. P. Mortenson, a field manager for Land O'Lakes, noted that at the Rice Creamery, "the Operator has always said plainly that the patrons would not stand for grading very closely."[32] In such cases, buttermakers worried that if they insisted on higher-quality cream, their patrons would take their cream someplace else and convince their neighbors to do the same. Farmers disenchanted with buttermakers' actions had ample opportunity to air their complaints and gain like-minded allies as farmers congregated outside the local creamery at delivery time. Patrons' pressure had especially strong influence in the late 1910s and early 1920s, when competition among creameries was at its peak.

By the mid-1910s, in specialized dairy districts, a growing number of creameries and cream stations competed to obtain farm families' cream.[33] Facing stiff competition from centralizers, buttermakers were especially reluctant to hold farm people to high sanitary standards. W. W. Clark, the agricultural extension agent in Houston County, Minnesota, noted that only one of the twelve creameries operating in the county did not directly compete with the territory of a centralized creamery.[34] As creameries struggled to get enough cream to keep their factories operating, buttermakers worried as much about getting enough cream as they did about obtaining high-quality cream. Some creameries even based buttermakers' salaries on the number of patrons they attracted, so it is not surprising that some buttermakers accepted cream of lesser quality. The success of small cooperative creameries rested on amicable relationships between the buttermaker and the farm people who supplied them.[35] Thus, although hand-cranked separators quickened the pace of cream separation and enlarged the scale of butter manufacture, they intensified problems with crafting a sanitary, standard product.[36]

Creameries' efforts to get good cream became even more difficult by the 1920s and 1930s, as motor-trucks and the improvement of country roads

hastened travel times in the countryside. In theory, such technological developments facilitated quick cream delivery.[37] But like the cream separator, the automobile had mixed effects on butter quality. As automobiles quickened the pace
of transport, crossroads creameries that once collected cream only from nearby
farmers extended their reach. After the arrival of the automobile, crossroads
creameries faced pressure not just from centralizer creameries, but from other
small factories once thought too distant to be a convenient market.[38] Even as
farm separators and motor trucks sped the trip from farm to factory, socially
agreed-upon standards of purity commanded great power.

Purifying Cream: A Social, Environmental, and Political Process

If buttermakers were to produce butter of consistently high quality and sanitation, they needed to improve the quality of cream they received. Buttermakers
did not seek to eliminate all bacteria from milk; in fact, most used bacterial
starter cultures to "ripen" cream before churning. The problem was not with sour
milk per se but with cream that had soured from abnormal ferments instead of
beneficial bacteria. Cream that arrived at the creamery clean and uncontami-
nated could ripen and produce good-flavored butter, but poor cream carried
bacteria that marred its flavor. The variety and amount of acidity in cream varied
with the age of the cream, the kind of feed cows ate, the phase of a cow's lactation, and the season of year.[39]

To overcome this natural variability, factory managers proposed two kinds of
solutions. The first alternative that creamery operators utilized to control sour
cream was called *neutralization*. Buttermakers tested the acidity of the cream
they received and then added a lime-water mixture to cream in proportion to the
cream's acidity. The more sour the cream had become, the more lime water had
to be incorporated to bring the acidity in check. Buttermakers aimed for a target
of 0.25% before churning it into butter.[40] Often buttermakers used cream neutralization and cream ripening in tandem to direct the souring process. Cream
might arrive at the creamery sour, be neutralized to reduce its acidity, pasteurized, and then be soured again with bacterial cultures before being churned into
butter.[41]

A second technique, taken up by increasing numbers of buttermakers in
the 1920s, was to make butter from sweet cream—cream with a lactic acid
content of 0.25% or less. Sweet cream butter often received a higher score from
butter testers and commanded a higher price.[42] Some consumers preferred its
milder taste.[43] Sweet cream butter's most important attribute was its keeping
quality. Butter made from ripened cream deteriorated more quickly than that

made from sweet cream and developed fishy, oily, and metallic flavors over time. As nearly 25 percent of the nation's butter was stored to be sold during the winter months, a long shelf life was important.[44]

The debate over how to overcome problems with spoiled cream paralleled earlier discussions about how best to purify milk. Like pasteurization, neutralization was a technical fix to be carried out by the dairy manufacturer. Like tuberculin testing, using sweet cream for butter production placed the onus of cream improvement on dairy farm families. Proponents of neutralization, like supporters of pasteurization, argued that it was cost-efficient and eliminated waste.[45] Those who opposed neutralization, like those who opposed pasteurization, worried that the process would discourage improvements to the sanitary conditions in which cream was produced. If neutralization enabled soured cream to be made into salable butter, what incentive would farm families have to keep cream cool and clean on the farm?[46]

The method a creamery utilized to overcome spoilage and define cream's purity—whether it neutralized cream or obtained it sweet—indicated how it balanced the interests of farm families supplying the butter trade and the imperatives of an industrializing economy. Neutralized butter stood as a symbol of big business, because centralizer creameries pioneered the technique.[47] Small creameries, with their more direct contact with farm patrons, led the way in making sweet cream butter. Small creameries hoped that selling sweet cream butter would distinguish their product from that of centralizers on the national market and enhance farmers' profits.[48]

Those that produced sweet cream butter worked to tilt the regulations in the marketplace to their favor, proposing legislation that distinguished sweet cream butter from competitors' neutralized variety. One reform was to require that butter made from neutralized cream be labeled. Their hope was that consumers would believe that neutralized butter was an unhealthy or inferior product to butter made from sweet cream.[49] Sweet cream butter promoters also argued that neutralized butter was adulterated. Drawing on a 1902 federal law that established a tax for adulterated butter, they contended that neutralized butter should be taxed.[50] In 1920, the Commissioner of Internal Revenue declared butter made from cream high in acid (including cream that was neutralized) was adulterated butter and therefore subject to tax.[51] The tax, at the rate of ten cents per pound, provoked quick response from creamery manufacturers that practiced neutralization and was reversed in September 1921.[52]

As federal regulations waned, state food and dairy inspectors aided sweet cream butter proponents' efforts. Minnesota's State Dairy and Food Commissioner defended the measure that classified neutralized butter adulterated at a hearing on the treasury ruling.[53] In Michigan, creameries that demonstrated that they could meet high sanitary standards carried a special state label and sold butter at a higher price. The Dairy and Food department provided a list of names and addresses using the state brand to interested consumers.

Despite winning state sanction, the efforts of small cooperatives to distinguish sweet cream butter from counterparts had limited impact. James Helme, the former commissioner of the Michigan State Food and Dairy Department who devised the state brand for butter, acknowledged that "it may take some time before Michigan State brand butter is known in the markets of the great cities."[54] In neighboring Wisconsin, when 306 housewives were asked "what does sweet-cream butter mean to you?" in 1932, half of the respondents had no response. Some women identified sweet cream butter as high-quality butter, but a few commented that sweet cream butter had less flavor, a flat taste, and too pale a color. Two women even contended that "you can't make butter from sweet cream."[55] As centralizer and local creameries fought over cream neutralization, consumers—even in a dairying state—lacked a clear understanding of what the terms *sweet cream* and *neutralized cream butter* even meant.

At first glance, neutralization seems to offer the most dramatic example of dairy industrialization during this period. Endorsed by large centralizer creameries, neutralization accommodated the goal of cream purity to the economy of scale. But manufacturing sweet cream butter was just as revolutionary and far-reaching a transformation to the butter trade. Implementing it required changes to established rhythms of farm work, new networks of product distribution, and different ways of thinking about the nature of the dairy farm. Like neutralization, the shift to sweet cream butter was a modern phenomenon.

The Nature of Mass-Marketed Butter

Creameries that sought to make sweet cream butter had to convince three audiences of its benefits: farmers supplying cream, state agencies that monitored the boundaries of purity, and consumers who purchased butter. The most profound change required was to convince farmers to devote the time to dairy work that selling sweet cream required. Most farm families conceived of cream as merely one element of a multifaceted farm income stream, not as a central economic engine. Many farmers spent more keeping a cow than they received for the butterfat it produced each year. They valued cows not simply by how much the animals earned in cash receipts from cream sales, but also for the skim milk, manure, and dairy calves they produced. Such farmers did not operate without regard to cash revenues; rather, they valued the cash garnered from selling corn fertilized with manure, or hogs that had been fed skim milk, both byproducts of dairying, like cream itself.

To reformers, this method of dairying had limitations. As A. J. McGuire, the general manager of Land O'Lakes Creameries, noted in a 1919 speech to Minnesota farmers, "Dairying in this way [for manure and skim milk] pays because it keeps up the fertility of the land, but the average young farmer will not become very enthusiastic over his dairy work unless he can be shown that there is something more than skim milk and manure for his labor in dairying."[56]

The more to which McGuire alluded was farm profit, and if farmers reorganized their farms to maximize cream production, McGuire believed, they could rely on dairy cows' cream, not just their byproducts, as a profit-builder. McGuire emphasized the need for cash profits to keep the *young* farmer satisfied. McGuire delivered his speech at a time when rural people and policymakers were especially concerned about keeping young women and men from leaving the farm. Only by making the farm more akin to the world of industry, with regular forty-hour weeks, modern machinery, and a steady paycheck, he argued, could rural areas keep the youth eager to convert their labor into the goods of consumer culture.[57] Farmers could partake fully in the industrial system, McGuire believed, but only if all of the goods they produced garnered high prices in the national market.

In the 1910s and early 1920s, only a few farmers conceived of the dairy work of their farms in the same way as McGuire. Many farmers were willing to feed cows at an economic loss for their manure and skim milk. Manure restored nitrogen, phosphorous, and moisture to soils without requiring outlays for commercial fertilizer. Thus, even after synthetic fertilizers became available, farmers looked upon manure favorably as a home-grown soil improver.[58] Collecting dairy cows' manure frequently generated higher yields of farm crops and improved barnyard sanitation.[59] Thus, hauling manure was one of the farm tasks that Elihu Gifford recorded most frequently in his diary, second only to cutting wood. (Ordinary daily tasks, such as feeding livestock or milking cows, did not even merit mention in the diary's pages.) Gifford hauled manure most frequently in fall, winter, and spring, when cows were being fed in the barnyard and when caring for crops didn't consume most of his time.[60] Similarly, skim milk boosted farmers' return from hogs and calves. Gifford valued skim milk so much that in 1911 he purchased eight cans of skim milk at ten cents each.[61] Wisconsin farmers Margaret and Rangnar Segerstrom likewise remembered raising "really nice pigs and calves" on skim milk from the cream separator.[62] Whereas farm people like Gifford and the Segerstroms saw dairy work tightly intertwined with other farm operations, experts sought to loosen its connections to hog- and crop-raising and elevate the place of dairy work on the farm.

A second element of dairy experts' market-driven reform effort was to encourage farm families to even out the seasonality of milk production with a balanced ration so they could take advantage of higher off-season milk and cream prices. Generally, cows gave the most milk during the late spring and early summer, when they grazed on pasture grasses and calved. Cows thrived on well-tended pasture grass, for it was high in nutrients, succulent, and tasty. But the seasonal increase in milk production that grazing encouraged, rooted in the ecology of the seasons, had an acute economic impact on milk prices. During the flush summer months, milk and cream prices fell with the abundance of milk produced, while milk prices increased in the winter due to the paucity of supply. The seasonality of milk production also proved frustrating for many farmers, making the burden of dairy work most intense in the summer, when crop work also demanded their attention.

Farm experts wanted farmers to provide cows a ration to encourage high milk production year-round. Ideally, greater attention to animal feeding would also more equally distribute the burden of dairy work throughout the year. Dairying experts' feeding recommendations, rooted in market demands for a reliable and consistent quantity of cream for manufacture, introduced different uses of pastures and mixes of field crops to local farms. The recommendations provide a key example of the way that agricultural industrialization reordered ecological, as well as economic, relationships.

Dairy rations were composed of two parts: roughage (hays, straw, silage, or corn stover) and concentrates (grains or meals). Concentrates were usually the most expensive part of the ration. Farmers could rarely grow enough grain to feed cattle and had to purchase it from feed dealers. The more nutrients cows received from hay or silage, the less high-priced concentrates farmers would need to supply. Thus, farmers sought feed rations that minimized the need to buy expensive concentrates. Franklin Pope Wilson, who operated a farm in Loudon County, Virginia, for instance, wrote to the United States Agricultural Department's dairy division in 1923, with hopes of devising a ration with a greater proportion of home-raised feeds. "With the present high prices of concentrates... bran, cottonseed meal, and mixed feeds, and the low price of wheat, which we raise and of which we have a supply on hand, I would like to know whether we can use to advantage some of this low priced wheat to mix with crushed corn, which we also raise, and with a more limited amount of the high priced bran and cottonseed meal which we have to buy."[63] Dairying experts' most frequent advice to farmers was to improve the quality of the roughage they fed, especially by furnishing cows with corn silage and high-protein hays.[64]

Corn silage was one roughage that dairying experts promoted. Made from the cut-up stems and leaves of corn plants, cows relished silage because of its succulence. Dairying experts liked it because it extended the milking season and boosted milk production.[65] Fed alongside hay, silage whet cows' appetite for other foods and provided them with extra nutrients for milk production.[66] Storing silage enabled farmers to have an ample reserve of fodder for winter months and also for midsummer when pasture grasses dried up and failed to provide a good ration.[67] A filled silo reduced the risk of price fluctuations or intemperate climate by ensuring farmers a ready part of cows' rations.

Corn had a long history as a livestock feed, but its use as silage was still relatively new to dairying in the early twentieth century. Some feared its effects on animal health, while others worried about silage flavors in milk.[68] In 1917, only approximately 11 percent of dairy farms had silos.[69] To take advantage of the risk-reducing and nutrient-improving potential that silage provided, farm families had to carry out a number of additional tasks: to construct a silo, to cut silage, and to fill the silo each September, in addition to raising a corn crop. When Elihu Gifford constructed a new silo in 1916, its assembly consumed much of the spring. The parts of the silo arrived by train in February, and he drew them in multiple sleigh-loads between the train car and the farm. Then, in early June, he dug a foundation and hewed scaffold poles in preparation to build the silo. With the help of three neighbors, Gifford and his brother began construction. Not until a month later was the thirty-three-foot silo fully tiled, roofed, and shingled.[70] Gifford, who enjoyed community members' assistance in building his silo, helped his neighbors to fill their silos each fall in return. Gifford or one of his sons brought their engine to power a neighbor's silage cutter, which sped the process. To see a silo on a farm signaled the farm's orientation to the tenets of industrial farming; farms equipped with them yielded greater quantities of milk more efficiently that could be channeled to a national market. But the neighborhood labor that constructed and filled silos each year is a reminder of the continued local character of industrializing dairy farms in the 1910s and 1920s.

Alfalfa and clover hay were other feedstuffs that buttermakers thought would make for a more profitable cream trade. Whereas corn provided cows' digestible carbohydrates, clover hay and alfalfa hay were both high in protein. Together, the crops balanced out the ration and minimized the need for protein-rich concentrates.[71] Red clover had additional virtues. It added nitrogen to the soil and was hardy enough to thrive in all kinds of weather. As Arthur McGuire told a gathering in 1929, "We would like to have every Land O'Lakes Creamery patron know and get the value of sweet clover pasture. It is without question twice the value of any other pasture. It comes on earlier in the spring, grows faster, withstands dry weather, and grows until the ground freezes in the fall."[72]

Finally, dairy reformers urged farm people to take better care of cream. Carrying out a higher standard of cleanliness on the farm, they promised, would reap economic rewards through higher prices for sweet cream butter. They asked farmers to keep cream cool and to deliver it frequently. By the late 1920s, electric refrigerators came on the market, but not until the late 1930s would many rural districts acquire electric lines to power cream coolers.[73] Until then, farmers harnessed the powers of wind, ice, and cool streams to chill cream. The most common method was to submerge cream cans in a tank of cold water. In Minnesota, Land O'Lakes encouraged farmers to construct windmills to pump water directly into cream-cooling tanks. North Dakota dairy experts experimented with ice well refrigeration.[74] North Carolina farm families submerged cream cans in a barrel and set them in nearby springs to keep the cream cool.[75]

Modernizing creamery managers also asked farmers to deliver cream frequently or early in the day, especially in the summer, since the heat accelerated spoilage.[76] The Hillpoint Creamery Association, for instance, stated in its bylaws that cream be delivered to the creamery by 8 a.m. between May and October, and by 9 a.m. the rest of the year.[77] Most Wisconsin creameries surveyed in 1932 received cream three times a week in the summer and twice a week in the winter, but some required more frequent deliveries. Members of the Land O'Lakes creamery cooperative had to deliver cream four times a week during the summer and three in winter, and some creameries even called for daily deliveries during hot weather.[78] Earlier summertime deliveries cut spoilage by minimizing time from cow to creamery and also gave buttermakers more time to handle the greater volume of cream produced during the flush season of May, June, and July.

Creameries had a particularly tough job convincing farmers to take on the additional work to keep cream sweet. For farm people, frequent cream delivery meant spending more time on dairy work during the busy season of planting and cultivating corn, mowing hay, and cutting oats. To encourage farmers' efforts, buttermakers graded cream at the factory and rejected cream that they deemed unfit for making butter, while paying a premium for the highest-quality product.[79] In 1921, for instance, the Minnesota Cooperative Creameries Association began paying patrons an extra five cents for each pound of sweet cream they delivered.[80] But only a minority of creameries practiced cream grading. Only 28.6 percent of Wisconsin creameries surveyed in 1932, for instance, graded cream.[81] Of those who did, fewer still paid a differential rate that rewarded the best cream. More plants cut the price paid for the poorest cream received, thereby eliminating the worst cream and improving quality.[82]

When looking for evidence of industrialization, historians generally hone in on the adoption of new technologies or the expanding scale of farm operations. But the signs of farmers' incorporation into a modern, industrialized butter market were often more subtle: a cement cream-cooling tank filled with well-water or a field seeded in alfalfa hay. Such small-scale farm improvements indicated the power of a new national standard of purity to reform farm practice and a newfound desire to see dairy cows not just as producers of byproducts to be utilized on the farm but as creators of cream profits. The added labor such improvements entailed, farmers hoped, would be compensated with the earnings in an expanded market. Even as they aimed to purify the cream they churned in different ways than centralizers, operators of small local creameries were driven by the same aim of achieving a consistent, uniform product. Whether produced by local factories or highly capitalized centralizers, mixed from neutralized cream or sweet, butter was a product of nature, human technologies, ideas, and labor.

Butter and the Rise of the Self-Service Grocery

As dairying experts worked to alter the ways that farm people carried out their work, new practices of retailing and manufacturing prompted changes in the way people purchased butter. Until the 1910s and 1920s, many consumers bought butter in bulk, with the assistance of a shopkeeper. Grocers carried different kinds of butter in wooden tubs in refrigerated cabinets behind the store counter. The tubs usually bore some description of the place from which it came, though rarely the specific creamery. When a consumer came to purchase butter, the retailer scooped out a specified quantity from a butter tub, weighed it, and wrapped it for taking. Consumers relied on price, color, and storekeepers' recommendations to decide which butter to buy; once home, when they tasted its flavor, shoppers decided whether the grocers' word could be trusted in future purchases.[83]

By the interwar period, consumers increasingly bought packaged butter from the refrigerated case of a chain store. Suggestions of state-backed nutritionists informed purchasing decisions alongside grocers' words and product labels. As in the past, consumers knew butter by tasting and smelling it, but their perceptions of the food were also mediated through the labels it carried, the sites where they consumed it, and state policies that governed its price.[84] Changing practices of butter buying, like new forms of butter manufacture, carried implications for how people learned and drew the boundaries of pure and impure food.

Advertisements and food labels told consumers that butter's purity originated in one of two places: the pastoral landscape or the modern factory. Butter appeared either as an industrial product or a product of nature, but not one of an industrialized nature. Claims about butter's purity simultaneously obscured the processes of change transforming the countryside and the environmental connections inherent in the mass-produced food. In political debates about how to distinguish butter from oleomargarine, however, a different vision of butter surfaced, one that betrayed that its composition had both technological and natural origins.

As cooperatives and centralizer creameries battled over issues like neutralization, the primary concern of most consumers was butter's price. In the 1910s, food prices spiked. The average retail price of butter, just under forty cents in

1913, rose to nearly seventy cents per pound by 1920.[85] Many consumers blamed inefficiencies in the commodity chain between farm and market.[86] They felt that wholesalers and retailers unfairly raised the prices of household staples and that more efficient methods of distribution might eliminate these costs. Grocers received special condemnation. In the years before self-service, consumers depended on grocers to monitor quality and measure quantity of butter they received. If a shopper found that the butter provided was overly salty or tasted moldy, she held the grocer partly responsible for insufficiently screening the quality of goods provided. Bargain-conscious shoppers also worried that grocers weighed butter improperly, so that purchasers received less than their money's worth.[87] Consumers' dissatisfaction with full-service grocers in a time of rapid inflation created the context for the introduction of branded, packaged butter.

Like consumers, farm-based cream sellers and small creameries expressed frustrations about the wholesalers and distributors to whom they sold their cream. Each transaction between farm and market cut into the profits they received for the good. They resented paying transportation charges and worried that wholesalers marketing and storing their butter might not care for it properly. In some cases, consumers' and farmers' dissatisfaction with existing modes of food distribution yielded direct farm-to-market sales.[88] But the geography of butter production limited the utility of farmers' markets. Many creameries and cream sellers were located hundreds or even thousands of miles from the people who purchased their goods.

Instead of turning to direct marketing, butter buyers and cream sellers turned to the newest model of food marketing: the self-service chain store. The hope was that self-service grocery chains, by buying in quantity and cutting costs on service, would deliver goods more efficiently, allowing consumers to pay a lower price for a quality good and limiting the intermediate charges that minimized farmers' earnings. Chain store purchasing appeared to make butter buyers more independent and self-sufficient and less dependent on grocers' advice. In fact, it replaced grocers' role with advertisers and nutritional experts, enlarging the power and influence of retailers in guiding butter's path from farm to market.

Self-service food chains minimized the role of wholesalers. Instead, representatives of the chain obtained butter directly from the manufacturer. Creameries, including small cooperatives, signed up for chain store contracts because they believed that direct contracts would bring cost savings. As Philadelphia's American Stores Company promised Wisconsin's Milltown Co-op Creamery in 1920, "We deduct no commission, no cartage, nor any other incidental charges except freight." Working with a chain meant access to a steady, year-round market for small-scale creamery patrons. To meet the quantity demanded by their contract with American Stores and to save in shipping rates, managers of the Milltown creamery teamed up with creameries in Luck, Frederick, Centuria, and Dresser Junction, Wisconsin, before shipping carloads of butter en masse to Philadelphia and New York.[89]

As it circumvented wholesalers, the self-service chain also minimized the role of the grocer as mediator of each butter-purchasing transaction. Instead, consumers were to determine the merits of a product by relying on so-called silent salesmen: packages, advertisements, and product displays.[90] Proponents of the new sales method told business leaders that packaging made a sales pitch more effectively than people did. In the words of one 1928 packaging guide, "The manufacturer cannot expect the average retail clerk…to put up a very strong selling talk." A well-designed package, by contrast, provided manufacturers with control over the messages delivered about their product and reduced labor costs.[91]

Proponents of packaging also claimed that labels enabled butter manufacturers to differentiate their product from others offered in the market. As the Paterson Parchment Paper Company explained to buttermakers in 1915, by stamping one's name on the butter package, buttermakers could "create a demand with the housewife, so that when she comes to purchase butter she will look for, ask for, and insist upon having his particular brand." The company continued, "It is worth a great deal to you to have a characteristic package that cannot be forgotten."[92] In time, the Paterson Company's prediction rang true. One 1931 survey found respondents could name forty-eight different brands of butter.[93]

Before brand recognition could take hold, advertisers had to convince purchasers to select packaged rather than bulk butter. Butter advertisers tapped into consumers' dissatisfaction with full-service grocers to promote the packaged good. An advertisement for Wedgewood Creamery butter alluded to the potential for grocers' to cheat the scales by stressing that its package carried a *"full sixteen ounces."*[94] Packaged butter ads also suggested that bulk goods posed sanitary hazards. A 1911 Meadow Gold advertisement, for instance, noted that by wrapping its butter in extra layers of packaging, it remained free from grocery-store odors and woody flavors that tub butter absorbed.[95] In 1930, Land O'Lakes published a similar advertisement, contending that bulk butter was a "menace to health."[96] Criticisms of bulk butter faulted retailers, not manufacturers, for impure or unclean butter.

By contrast, butter labels and advertisements praised the sites where butter originated: pasturelands and butter factories. Multicolored labels depicted scenic vistas and grazing cows. They created brand names with natural connotations. In 1924, the Minnesota Cooperative Creameries Association's adopted the name Land O'Lakes, linking their butter to the streams and lakes of the place that produced it.[97] Even butter factories that constituted a part of urban meat-packing operations adopted natural-sounding names that belied their modernity; Armour called its butter "Cloverbloom" and Swift used the name "Brookfield."[98]

7

A Tub of Trouble

Wooden tubs attract mould and mould spoils butter. A disagreeable "woody" taste is also absorbed from the tub. Exposed butter, whether in tubs or prints, never retains its freshness and purity.

A Package of Purity

In making Meadow Gold Butter strict attention is given to cleanliness. From the pasture to the package scrupulous care is constantly exercised. Nothing harmful ever comes in contact with the milk or cream.

Meadow Gold Butter is a pure product from a model creamery. It is packed directly into airtight packages which preserve its fresh, delicious flavor and natural sweetness. *Ask your dealer for it.*

P. BERRY & SONS, Hartford, Conn.
Sole Distributors for New England.

Figure 2.2 The first step to branding butter was to discredit bulk butter as an unsanitary food. N. W. Ayer Advertising Agency Records, Archives Center, National Museum of American History, Smithsonian Institution.

The nature to which butter labels appealed was a timeless one. Advertisements depicted clear streams and grazing cows, not newly constructed cooling tanks or silos. The labels acknowledged that butter derived from herds of Jerseys and Holsteins, but made no mention of the farm people who milked the cows, separated the cream, or delivered it to the creamery. The Land O'Lakes label made an Indian maiden its central figure. By associating butter with these images, creameries assured consumers that even as the world around them changed, their butter remained authentic and simple.

At the same time as they praised pastoral landscapes, butter labels highlighted creameries as modern manufacturers. Meadow Gold told potential purchasers that each churning was "analyzed by skilled chemists." Advertisements emphasized that machinery, not human hands, prepared it for sale. Fairmont Creamery promised that each employee donned a "freshly laundered uniform each morning" and guaranteed that at each stage "expert specialists" oversaw creamery processes, using scientific techniques to ensure "uniform body, texture and quality."[99] Emphasizing butter's industrial

Figure 2.3 The vision of butter's purity expressed in the 1920s underscored the food's natural origins and the expertise of those who churned it. Note the emphasis to consumers on getting the full weight in a packaged good. N. W. Ayer Advertising Agency Records, Archives Center, National Museum of American History, Smithsonian Institution.

origins, these ads credited human expertise and technology for ensuring butter's quality and minimized the seasonal variability or generative powers of nature on which cream production depended.

In truth, butter was not wholly natural or technological, nor was its purity rooted simply in its pastoral origins or industrial safeguards. At every stage, butter was a hybrid of natural and cultural influences, from the moment that farm people led a cow to a bull for breeding to the time that consumers' bodies converted the dairy fat into energy. Advertisements juxtaposed representations of the natural pasture and the technological creamery without comment. But political debates about butter's place in the national diet brought the tensions such representations suggested to the fore.

Butter, Margarine, and the Nexus of Nature and Culture

Even as the rise of chain store retailers encouraged creameries to begin to brand and differentiate their butter from that of other manufacturers, what concerned many in the butter industry was not competition from other buttermakers, but competition posed to their product by oleomargarine. Dairy farm families and butter industry officials had lobbied against margarine since the 1880s, and their efforts had successfully created a federal margarine tax. But by the interwar period, competition between butter and margarine intensified as butter prices increased and new technologies altered the way the fats were made. Dairy interests consistently appealed to the *nature* of butter and disparaged margarine as an artificial food. Despite these claims, both butter and margarine were mixed products of environmental forces, technological interventions, and social mores.

One of the most important new developments in the competition between butter and margarine in the interwar period was the discovery of vitamins in foods. Vitamin research transformed the way in which Americans evaluated foods' health value. Once feared for its role in spreading communicable diseases, food was newly lauded for its beneficial properties. Further, foods like fruits and vegetables, discounted in a model of nutrition that emphasized high calories and low costs, came to be appreciated as "protective foods" that delivered essential nutrients.[100] As late as 1905, nutritionists with the United States Department of Agriculture advised consumers to purchase foods that provided the greatest amount of calories, fat, and protein for the lowest cost.[101] But by the late 1910s and 1920s, research indicated that certain fats and proteins carried beneficial trace nutrients, called vitamins, that warded off diseases like rickets, scurvy, and pellagra. The proper ratio of carbohydrates, proteins, and fats alone would not guarantee good health. Rather, a mixed diet of vitamin-rich foods was important. Getting enough vitamins required careful food selection, preparation, and consumption.

Butter manufacturers followed vitamin research closely and benefited from the publicity it generated.[102] When animal feeding experiments found that butter, unlike other fats, contained vitamin A, researchers suggested that butter was nutritionally superior to other fats.[103] Home economists, whose numbers grew and professionalized in the 1910s and 1920s, disseminated nutritional advice based on vitamin research to a wide audience. Public health officers also publicized findings about vitamins, enlarging their role as food inspectors and encouraging children to eat vitamin-rich foods like butter. Aided by public health officers and nutritionists, butter manufacturers advised consumers that the fat in butter was healthful by its nature.[104]

Vitamin research articulated the relationship between nature and health more narrowly than images of the pastoral landscape used to elicit purity. It established the specific pathways by which nutrients of nature entered human bodies and measured the material composition of foods. By the mid-1930s, for instance, researchers knew that cows fed roughage high in carotene—such as fresh alfalfa, carrots, or fresh green Kentucky bluegrass—converted the carotene pigments to vitamin A and secreted it in larger quantities in their milk.[105] The more directly a cow's diet came from nature, the higher the vitamin content of the butter produced from its milk. Rather than replace aesthetic appeals to nature with nutrition-based claims entirely, advertisers clothed old notions of the health of the country with the new legitimacy afforded by vitamin research. Fairmont Creamery Company, for instance, reminded consumers that "Fairmont's Better Butter Brings the Vital Food Element Vitamins from Fields of Clover and Alfalfa to your Table."[106] By tying its butter to clover and alfalfa fields, Fairmont was able to evoke the pastoral landscape and utilize scientific studies in one fell swoop. Even with new nutritional knowledge, the most basic advertising claim employed by buttermakers remained: butter was healthy by its nature.

In 1938, when scientists fortified margarine with vitamin A, they undercut the inherent nutritional advantage butter held. As vitamin fortification made margarine nutritionally equivalent to butter, dairy industry officials turned to well-worn rhetoric about nature to bolster butter's standing. Whereas fortified margarine's vitamin richness was counterfeit and artificial, they claimed, butter's nutritional content was "natural."[107] To the surprise and dismay of the dairy industry, nutritionists expressed few qualms about recommending vitamin-fortified margarine as a healthful alternative to butter.[108] A 1941 study by the American Medical Association's Council on Foods and Nutrition, for instance, concluded that "there is no scientific evidence to show that the use of fortified oleomargarine in an average adult diet would lead to nutritional difficulties."[109] The Committee on Food and Nutrition of the National Research Council similarly determined that "present available scientific evidence indicates that when fortified margarine is used in place of butter as a source of fat in a mixed diet no nutritional differences can be observed."[110] By 1942, 99 percent of all margarine was fortified with vitamin A.[111] The language of vitamin research, once used to describe butter's natural properties in modern terms, ultimately encouraged scientists and consumers to recommend a replacement.

Dairy industry officials invoked the connections between nature and health in a second element of the butter-margarine debate: disputes over butter's color. State and federal legislation regulating the sale of oleomargarine turned on whether margarine manufacturers should be allowed to color their product in the same golden shade as butter. In the 1880s, numerous states passed margarine laws prohibiting the coloring of margarine. New Hampshire, Vermont, and South Dakota even required that oleomargarine be colored pink so it would not be confused with butter. Color became the basis for federal laws to restrict oleo-margarine sales in 1902. In that year, a Congressional Act known as the Grout Bill increased the tax on "artificially colored" oleomargarine to ten cents per pound but lowered the tax on uncolored oleomargarine to one-fourth of one cent.[112]

Congressional laws taxing artificially colored oleomargarine and the dairy lobbyists who encouraged their passage drew a clear line between nature and artifice. According to butter proponents, butter's golden color was authentically natural—derived from the sun and the pasture grasses on which cows grazed. Butter's golden hue indicated its freshness and richness, a reminder of butter's pastoral origins.[113] As George McKay, a spokesman for the American Creamery Butter Manufacturers, declared in 1918, "In selling butter we might say we are selling air, sunshine, and rain."[114] Meanwhile, dairy lobbyists stated that the addition of yellow dyes made margarine artificial. They poked fun at margarine by claiming its origins were unnatural; while butter was the product of pastoral places, margarine came from an aberrant creature—a coconut cow.[115]

The line between nature and artifice was trickier, however, than the rhetorical stance of dairy lobbyists suggested. Butter's color was not, in fact, only derived from nature. During the winter months, when cows were fed on silage and dried hay instead of fresh grasses, their milk paled. Hence, butter manufacturers—like margarine makers—added yellow dyes to make butter's color richer.[116] More-over, margarine's color was not as artificial as rhetoric implied. In fact, by the late 1920s and 1930s, margarine manufacturers relied not on dyes, but on tropical oils to lend margarine a yellow tint. Combined with coconut oil, palm oil gave margarine a golden hue. Palm and coconut trees were, of course, natural species. Despite the natural origins of tropical oils, dairy lobbyists contended that mar-garine manufactured from them was artificial. According to lobbyists, the conscious selection of naturally colored yellow fats to make margarine was a manipulation of nature—a process "tantamount to using the product commonly known as artificial color."[117]

In November 1930, however, the U.S. Commissioner of Internal Revenue David Burnet revised the line between artifice and nature. Burnet argued that margarine containing unbleached palm oil would no longer be classified as artifi-cially colored since its color came from a natural substance. This reclassification meant the product derived from palm oils would be taxed only one-quarter cent per pound, not ten cents per pound.[118] The dairy industry responded quickly. In January 1931, Vermont's Congressman Elbert S. Brigham introduced a bill calling for any colored margarine—whether colored with dyes or naturally occurring yellow fats—to be taxed at ten cents per pound. Brigham's proposal left the tax on uncolored margarine low but restored the provisions previously established under the Grout bill to all colored margarine, rendering Commissioner Burnet's ruling moot. Brigham's bill became law on March 4, 1931.[119] Burnet's 1930 ruling that margarine colored by palm oil was a natural color forced dairy interests to reconsider descriptions of butter as a natural food and oleomargarine as an artificial one.

Despite these changes, imagery of nature continued to figure into the butter versus oleomargarine battle. Dairy interests played up the idyllic nature of the dairy farm and cast suspicion on palm and coconut oil by emphasizing the foreign labor and primitive processes used to produce it. On one side was the pastoral imagery of a herd of cows near a babbling brook in a springtime pasture. On the other, a tropical forest in a faraway place, where foreigners harvested coconuts and palm fruits. Tennessee Congressman Ewin Lamar Davis argued, "This is a contest between wholesome butter produced from American cows by American citizens on the one hand, and oleomargarines composed of the palm oil of Java and Sumatra and the coconut of the Pacific and South Sea Islands." Pennsylvania's Congressman Franklin Menges told of palm fruit chaff being "tramped by bare-footed natives" to extract the oil from which margarine was made.[120] By contrasting tropical natives in the palm oil trade to American dairy farmers, a group whose racial makeup was predominantly white, Menges and Davis tied butter's purity to racial ideologies.[121] Their critique also cast butter as a more technologically modern product than margarine, for barefoot palm fruit-trampers were a far cry from the expert, white-coated creamery workers high-lighted in butter ads. For Menges and Davis, as for creamery manufacturers who labeled butter with images of verdant pastures and up-to-date equipment, butter's purity rested in both its natural origins and its modern transformation.

Support from southerners, who had previously aligned themselves with the oleomargarine industry, was critical to the passage of the 1931 Brigham Bill. Southern support for the oleomargarine industry had rested in southerners' defense of the staple crops important to the South: cotton and peanuts. Peanut and cottonseed oils constituted some of margarine's key ingredients. As dairying developed in southern states, representatives from Mississippi, Tennessee, and Texas became more inclined to back their states' burgeoning dairy industries. But the bill's success also rested on the changing blend of oils used to manufacture oleomargarine. The Brigham Bill targeted palm and coconut oils for regulation, not cottonseed, soy, or peanut oils grown domestically. The growth of the southern dairy industry, combined with the rising amount of tropical oils used for oleomargarine, brought southern agricultural interests in line with dairy lobbyists to protect butter in the market.

National congressional support for butter was short-lived. In 1933, scientists mastered the process of hydrogenating fats, making it possible to use domestic products—such as cottonseed and soybean oils—as the main ingredients for margarine. Thus, dairy interests could no longer cast margarine as a foreign product. Congressional representatives from southern and even midwestern states who had aligned with dairy interests to pass the Brigham Bill became loyal once more to the soybean, livestock, and cotton producers in their own districts.[122] Once again, a new means of blending the nature and technology of food altered the position of butter vis-à-vis margarine.

Ultimately, World War II, not new technologies or taxation policies, changed American food habits regarding butter. At first, the war seemed a boon to dairy producers. Military contracts and lend-lease exports ensured creameries with a market. Americans employed in wartime industries had a greater appetite for dairy foods and the ability to purchase them. By 1942, however, policymakers worried that the country lacked enough fats and oils to supply wartime needs. The fight in the Pacific cut off access to imported oils. Such fats fueled the stomachs of soldiers and civilians and were used to make paint, soap, and propellant powders.[123] As wartime planning dominated the policy agenda, even pats of butter atop a stack of shortcakes seemed essential to the nation's defense.

In the early years of the war, as butter became more scarce, consumers went out of their way to obtain it. In February 1943, Portsmouth, New Hampshire, residents lined up in the pouring rain to await the arrival of a large shipment of butter at the local First National Store. Wartime diaries note butter purchases jubilantly, making clear that such purchases were not commonplace. As one resident wrote, "Ma got a whole lb of butter! First whole lb for a long time!"[124] Even as politicians and nutritionists cast margarine in more positive terms due to its inclusion of domestically produced ingredients and its enhanced nutritional content, consumers still preferred butter.

Facing scarcities of fats and oils, on March 23, 1943, the Office of Price Administration began rationing butter and margarine to prevent inflationary pricing. Rationing tightened the competition between butter and margarine, because consumers had to redeem more ration points to purchase a pound of butter than a pound of margarine. Yet, at first, consumers held to their preference for butter over margarine. Some citizens even cared enough about butter to complain to government officials when they learned of shipments of butter being sent for lend-lease purposes, voicing resentment that foreigners would receive American-made butter, when they themselves had not eaten butter for months.[125]

As butter became scarce, many consumers tried margarine for the first time. Portsmouth, New Hampshire, resident Louise Grant recorded her first use of margarine in an August 1943 diary entry, writing "Colored & salted & sugared some oleo & it was alright. Butter is so scarce & takes 10 points to get a pound that you can't afford to use it on corn, etc." By the end of September 1943, butter had become a rare treat, rather than a household staple, at the Grant household. "Butter is going up to 16 points and that would be a whole week's points for one person. We haven't bot [*sic*] any for a long while." By October 7, 1943, she wrote, "No butter of course but then we don't use it now."[126] This family's transformation in food practices from butter to margarine was paralleled around the country. If consumers purchased butter, they might have to forgo meat, whereas they could purchase margarine *and* keep putting meat on their tables. Hence, consumers who had once spurned margarine as a low-class food came to accept it and found the cheaper alternative for butter satisfying enough to spread on their toast or to coat their vegetables. Consumers' shift from butter to margarine during the wartime period illustrates the ways that state policy, as well as advertising rhetoric, mediated consumers' purchasing practices.

Wartime milk shortages, not simply those of fats, made federal policy on butter subject to policymakers' scrutiny. By 1943, New York, Washington, D.C., and San Francisco faced temporary fluid milk shortages. Margarine supporters argued that by replacing butter with margarine, dairies could channel their products to fluid milk, rather than separating milk to make butter. In a House Agricultural Committee hearing in 1943, Texas U.S. Representative W. R. Poage remarked, "Is not the sensible thing for us to do during a period of time when we have a physical shortage of the products of the cow to replace that one product that can be replaced with a vegetable oil rather than attempt to convert the fluid milk into a product for which we have a palatable and suitable substitute?"[127] Policymakers considered milk irreplaceable, but they increasingly defined butter as a manufactured food for which there were good alternatives. By the mid-1940s, many consumers, retailers, and nutritionists, drawing on the notion of equivalent food values, saw butter and margarine as interchangeable.

The logic of nutritional substitution, however, obscured that butter and margarine originated from very different rural environments. Poage's suggestion that milk be channeled toward drinking rather than butter production overlooked the distinctions between farms selling cream and those marketing fluid milk. Most farm families who supplied butter factories could not easily convert to fluid milk production. They were too far from cities to market whole milk and usually lacked the cooling equipment and licenses to do so. A dairy could shift to supplying milk for drinking only insofar as the farms supplying the milk company were equipped with the labor and technology to meet stringent sanitary standards. These tight connections between dairy manufacturers and their rural suppliers made a seemingly simple substitution of one raw material with another much more complex than policymakers suggested. As butter's exclusive claims to nutritional richness and natural purity melted away, it created new challenges both for creameries and the producers who supplied them.

By the 1940s and 1950s, the dairy industry mounted a vociferous defense of butter. To contemporary eyes, the documents from this period seem laughable. Kitschy pro-butter advertisements glorified the 1950s kitchen. Some observers even decried oleomargarine as a communist plot or a tool of Satan.[128] Viewed in the context of changes to dairy manufacturing in the 1920s and 1930s, however, these zany pro-butter documents take on a new light. The farm people and creamery manufacturers who doggedly defended butter did so because they became devoted to cream production for profit, rather than simply as a sideline of general farming in the 1920s and 1930s. The very same pro-butter proponents who appear in the 1950s and 1960s to be agricultural fundamentalists, committed to an industry and food of the past, were on the forefront of modernizing the dairy industry just decades earlier. The margarine-butter battles of the 1950s and 1960s indicate more than simple economic self-interest. They also reveal growing disillusionment among those who hoped tighter links between the goods of the country and the urban marketplace would bring a brighter agrarian future.

It is ironic that butter producers sought to distinguish their product as an authentically natural food and to characterize margarine as an artificial, factory-produced one. By the interwar period, butter producers were just as reliant as margarinemakers on industrial techniques and mass-marketing strategies to sell their food. Specialist buttermakers standardized butter quality by practicing neutralization or being exacting about the kind of cream their plant accepted and employed artificial dyes to give butter a uniform golden hue throughout the year. Whether farmers sent cream to centralizer creameries or to neighborhood cooperative creameries, the butter trade operated with a greater scale and geographic reach than ever before. Processes of industrialization reached back to the farm itself, where farm people altered their work patterns to deliver sweet cream, erected new technologies (like silos), and planted different forage crops to prolong milk production. Modernization of the butter trade brought business-minded efficiency to the barnyard.

Farm people's and buttermakers' increasing reliance on technologies hardly indicated that environmental factors no longer affected butter production. Rather, the widespread use of technological processes came about precisely because environmental factors posed perpetual challenges to the food's commodification. Neutralization overcame natural variability of cream. Stored silage moderated the seasonality of milk production. Butter color imbued the dairy fat with a deeper hue in the winter months. Neither unadulterated nature nor wholly artificial, butter was a mix of human technologies and environmental forces.

As the butter trade blended nature and technology, it incorporated small-scale dairy farm laborers into networks of national and global commerce. Whereas present-day rhetoric stresses the existence of two agricultural systems—one local and self-sustaining and the other industrial and large-scale—cream-supplying farmers of the interwar period operated on the local and national levels simultaneously. The centralized factory system of butter manufacture provided a market to the smallest, least efficient cream sellers. Locally-rooted social and ecological networks sustained and fostered engagement in a national economy. Neighbors joined together to haul scientifically balanced feed supplements shipped from afar as well as to harvest locally produced hay. Rather than seeing the rise of national manufacturing and a mass market as forces that eroded local rural economic vitality, many farm families viewed engagement with national markets as a path toward (not away from) economic independence in the 1910s and 1920s.

As farm people and creamery manufacturers relied on new technologies to transform the material nature of butter, they articulated an ideal of food purity that drew upon both industrial and environmental values. Advertisers stressed either the food's origins on the pasture or its expert completion in the factory, even though achieving pure and flavorful butter depended just as much on intermediary processes such as cream storage on the farm or sufficient sanitation of cream separators. Definitions of butter purity that centered on middle steps on the commodity chain—such as "sweet cream butter" and "neutralized butter"—failed to resonate with consumers. Even though food purity remained important for those manufacturing butter, the emergence of new ways of thinking about butter, such as nutritionists' emphasis on vitamins and state policy that established ration points, encouraged consumers to consider butter's nutritional contents and its price, not just its purity.

In 1911, butter was considered such an essential, irreplaceable, nutritious staple food that a hike in the butter price inspired a protest movement. By the late 1940s, butter was increasingly thought of as one of a few comparable alternatives for enriching the table. The same kind of women who had once defended access to butter as a citizens' right—consumer advocates and home economists—testified in the 1943 and 1949 congressional hearings to encourage an end to the federal tax on oleomargarine. Despite butter's well-established place in American cookery and rhetoric that consistently trumpeted butter's natural advantages, margarine benefited from the development of vitamin fortification, new consumer practices during World War II, and the elimination of federal taxes on nondairy fats. State policy, particularly on the federal level, facilitated consumers' changing attitudes about the dairy food.

Mass-marketed, packaged, standardized butter was just one product of the industrialization of the dairy business. The same factories that made dairy foods like butter and cheese generated a stream of cast-off effluent in the form of skim milk and whey. These byproducts of dairying, which had once been incorporated back into farm processes as animal feeds, concentrated as the scale of industrialization increased. Dairy plant managers hoped to recapture some value from these byproducts by remaking them, too, for a modern marketplace.

Purer Streams and Predictable Profits

Dairy Waste in the Mid-Twentieth Century

Declining use of butter was only one problem facing the dairy industry in the middle decades of the twentieth century. Another problem plagued the streams near dairy plants. By the 1920s, the streams in rural areas with cheese factories and creameries festered with rotting dairy discharge. Excess buttermilk spilled from creamery spouts onto adjacent fields and directly into waterways. Ditches brimmed with whey, attracting flies and creating a stinky mess. Many creameries had no access to a city sewer. In 1930, only 17 percent of waste released by Wisconsin milk plants was treated, leaving 42,513 pounds of untreated whey, skim milk, and other byproducts to be drained into the state's waterways.[1] In 1940, buttermaking expert Otto F. Hunziker concluded that "numerous methods have been devised and tried" to treat creamery waste, but "all of them involve considerable expense…and few, if any, have proved entirely satisfactory."[2]

The release of skim milk and whey into streams was not a new practice, but changes in the rural economy of the 1920s and 1930s—the rise of automobile tourism and the farm depression—changed the way that industrialists understood and reacted to milk plant pollution. First, the increase in rural tourism during the interwar period made it more difficult for dairy plants to hide the deleterious effects of dairy manufacturing from the public eye. As the number of Americans with automobiles increased and rural roads improved, motorists sought to take invigorating outdoor adventures in the countryside.[3] When tourists who ventured to the countryside to be energized by fresh breezes found themselves breathing the foul vapors of "dairy air" instead, disappointment ensued.[4]

Dairy plant managers understood all too well the potential effects of such visits on the rural traveler. As one manager explained in the late 1920s:

> The automobile has brought the cheese factory odor and conditions to
> the attention of countless thousands of people, and no amount of

argument or favorable advertising is going to convince the casual traveler that our cheese is of a high quality, when the attending conditions
are such as to produce a feeling bordering on nausea.[5]

That industrialists worried that the odors emitted by spoiling dairy wastes would
make consumers leery of dairy products reveals that Americans continued to
draw connections between healthy products and healthful places, even after
germ theory and the emphasis on vitamins encouraged consumers to narrow the
criteria by which they defined healthful.[6] Dairy plant managers feared that tourists' exposure to the unsavory odors of dairy manufacturing would lead them to
rethink their purchases of dairy foods, despite product labels that trumpeted
their nutritional contents.

The dairy industry's long-standing use of natural scenes and ideals to promote the healthfulness of the product made the degraded landscape surrounding cheese and butter plants especially problematic. In the nineteenth century,
dairies sold "country milk" by claiming that healthy foods came from healthy
landscapes. By the 1920s, they designed butter labels that lauded the product's
natural origins as the source of its nutritional benefits. But when consumers
drove into the countryside, they saw the nature from which dairy products as full
of sanitary ills. Dairy waste undermined the pastoral iconography used to sell
dairy foods.

A second problem with the residue was that it appeared particularly wasteful
during a time of economic strife. Like soils scarred with erosion, ditches brim-
ming with whey and buttermilk signaled the ecologic ills and economic imbal-
ances threatening national prosperity. As skim milk and whey poured into waterways, the dairy economy in the 1920s and 1930s suffered from drought,
decreased consumption, and dropping commodity prices. In 1930 and 1933,
droughts damaged feed crops and pastures in the South and the West, boosting
feed costs and reducing farm profits. Milk surpluses mounted as consumers
limited their purchases of butter, milk, and cheese, and as farmers from other
agricultural sectors took up dairy farming to weather economic volatility in beef,
grain, and hog prices. Producers who had received an average of $2.50 per hun-
dred pounds of milk in 1924 received only $1.29 for the same amount in 1933.[7]

Tourism, combined with a new attention to wasted resources in the farm
depression, recast a persistent problem with dairy waste as one of national
economic significance. Before the interwar period, dairy wastes were largely addressed as local problems. Since creameries and cheese factories were rural
industries, it was most often farm families who suffered when whey and skim
milk flowed over their properties from dairy plants. In some cases, when dairy
wastes tainted waterways, landowners filed nuisance lawsuits.[8] Cheese and buttermakers who lacked reliable waste disposal systems attempted to resolve dairy

byproducts problems by sending waste home with farm patrons to use as animal feed.[9] By the interwar years, dairy plants began to consider whether they might be able to market dairy residues nationally or even internationally.

To this end, from the 1920s to the 1950s, managers of cheese plants and butter factories sought to make the substances left over from the cheese and buttermaking processes into consumer goods or industrial materials. Their efforts to transform whey, skim milk, and buttermilk were both technological and conceptual. One important facet of their work was to identify technologies to prepare byproducts for industrial applications. No less important was remaking the perception of dairy detritus. Rather than viewing whey, buttermilk, skim milk, and milk protein as wasted parts, dairy managers wanted industrialists and consumers to approach them as desirable products. The dairy industry was not the only industry to recover waste products for use in manufacturing, but it followed a different path than many other industries because health and World War II were so critical in facilitating byproducts' entry into commercial channels.[10]

Beliefs in milk's healthfulness aided and abetted efforts to remake dairy byproducts. In the 1920s and 1930s, when chemists sought to transform byproducts into raw materials for industry, critics argued that vitamin-rich foods like skim milk and whey should be used to sustain human lives, not to fuel industrial production. Attuned to critiques of applying dairy foods to industrial uses, manufacturers recast dairy byproducts as health foods in the 1950s. Skim milk's low fat content helped the product gain a newfound appeal among consumers. World War II also played a key role in this transition, a time during which skim milk powder became a staple for army rations and was used in relief feeding. As government contracts boosted farm income, farmers became convinced of the economic value of selling skim milk, rather than using it on the farm. New processes and technologies were important to solving the byproducts problem, but fresh perceptions about dairy byproducts as healthful sources of nutrients or as essential to the war effort were equally significant to the substances' transformation.

As technological innovations made wasted byproducts valued goods, they also rendered once-valued goods as waste. Such a development took place on dairy farms as new processes of dairy breeding made most dairy bulls redundant. By the 1940s and 1950s, rather than keep a bull to fertilize one's cows, some farm families turned to artificial insemination (AI). Farmers found that utilizing sperm from the best bulls of the breed improved their chances of boosting cows' productivity. Further, carefully managing the insemination of one's cows reduced the risks of livestock disease. As farmers relied more heavily on sperm acquired from studs in breeding cooperatives, the herd bull headed to the chopping block as a byproduct. At the very moment that manufacturers' decisions about dairy wastes altered the quality of rural streams, farmers' process of defining waste reconfigured the composition of the dairy herd.

The introduction of AI in the 1930s and 1940s brought the gospel of efficiency that transformed waterways and forests in the Progressive Era to the barnyard, and in so doing reconfigured farm families' relationship to individual animals. The imperative to make barnyard nature more efficient classified some animals (those with impressive milk-giving records) as indispensable and implicated others (bulls with less impressive records) as wasted resources. Historians have done much to investigate the implications of the application of the utilitarian principles of "greatest good to greatest number for the longest time" on communities affected by the erection of dams or the establishment of national forests. But rarely have they examined how devotion to such principles remade the ways in which farm families managed and related to domesticated animals under their care. Revisiting the story of dairy byproducts thus offers new perspectives on how and why industrial agriculture and consumer culture altered the way in which Americans valued nature in the middle decades of the twentieth century.

From Industrial Byproduct to Modern Marvel: Chemurgy and Casein

In the 1920s and 1930s, manufacturers first used organic chemistry to remake excess whey, skim milk, and buttermilk into commercial industrial products, hoping to increase returns to farm patrons and eliminate the sanitary and legal problems such wastes generated.[11] The transformation of agricultural byproducts into industrial resources came to be known as "chemurgy," a word coined by an organic chemist with Dow Chemical Company that paired the Egyptian "chemi" for chemistry and the Greek word "ergon" for work.[12] By 1935, three hundred scientists, industrialists, and members of farm organizations gathered to form the Farm Chemurgic Council, a group devoted to studying and establishing new nonfood uses for farm crops and markets for agricultural byproducts.[13] Unlike nutritionists, who lauded the balance of proteins, carbohydrates, and fats in milk, chemurgists sought to break down milk into its component parts and find uses for each of them.

The vision of chemurgy proponents revealed much about cultural attitudes of the 1920s and 1930s. Turning farm byproducts into modern industrial products helped reconcile the growing tension between modernity and agrarian tradition.[14] Utilizing dairy byproducts to fuel industrial development incorporated rural people into the promise of a modern industrial society and enabled them to afford the technologies it promised, without abandoning the countryside for the city. Chemurgy blended the commitment to an agrarian past and industrial

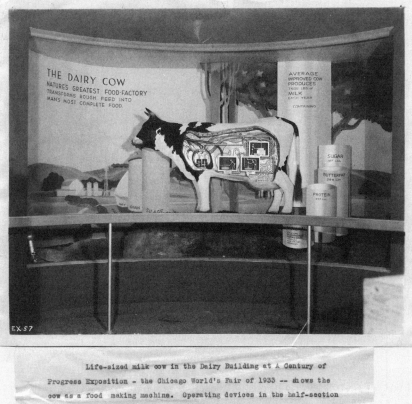

Figure 3.1 This display of the dairy cow at the Chicago Century of Progress exposition conveyed the idea of the cow as a feed-converting machine, one that transformed raw materials into marketable products. Science Service Photograph Collection, Division of Work and Industry, National Museum of American History, Smithsonian Institution.

future; it was a vision of modernity rooted in the continued economic viability of the farm. By basing industrial processes in nature—and especially in biological products that were renewable—scientists, industry leaders, and agricultural officials believed they could ensure a firmer foundation for the economy than the volatility of the stock market.[15]

Pairing efforts to use resources efficiently with those to improve rural living standards, chemurgists followed the broad outlines of what historian Sarah Phillips calls "New Conservation."[16] Like New Deal conservation programs that combined economic and environmental objectives by rebuilding soils and

Figure 3.2 This 1922 exhibit by the Dairy Division of the United States Department of Agriculture promoted the use of whey. Wisconsin Historical Society, WHi 93305.

simultaneously restoring farmers' balance sheets or hiring unemployed men as trail builders and tree planters, chemurgists believed they could revitalize the rural economy by making nature the foundation for modern life.[17]

In the dairy industry, chemurgists were especially interested in the protein of milk, called *casein*. Extracted from skim milk, the byproduct of butter processing, casein could be integrated into a wide variety of household goods. As early as 1900, Henry E. Alvord reported that casein could substitute for eggs, replace glue in paper sizing, and be hardened into plastic to make buttons, combs, or electrical insulators.[18] During World War I, casein coated airplane wings.[19] By the early 1940s, artwork created with casein paints graced gallery walls, theater stars sported dresses spun from casein fibers, and pianists tickled keys melded from casein instead of ivory.[20]

No matter how many ways chemurgists found to exploit casein's malleability in the marketplace, their efforts were often stymied by the cost of casein manufacture and the value of skim milk for other uses. One challenge for casein manufacturers was obtaining enough skim milk for its production. If skim milk were to be used for casein manufacture, it would have to be wrestled away from farmers who valued its use on the farm, for it took thirty-three pounds of skim milk to produce just one pound of casein.[21] Although skim milk flowed in the ditches of some manufacturing milk districts, farmers battled with factory managers over the fate of skim milk in hog- and chicken-raising districts because farmers wanted to use it as animal feed.[22]

By the 1930s, butter factories collected whole milk at the factory, making it possible to process dairy byproducts as well as to make butter. But farm families were still reluctant to allow creameries to keep skim milk and use it for industrial processes. Farmers expected to collect a supply of whey or skim milk in proportion to the whole milk they delivered.[23] Buttermakers had to apportion skim milk carefully to creamery patrons who wanted to use it on their farms. According to Floyd Lucia, a buttermaker who worked in a creamery near Green Bay, Wisconsin, "We had great tussles with the farmers, 'cause they'd take more skim milk than belonged to 'em!"[24] Throughout the 1920s and 1930s, animal feeding experiments demonstrated that livestock fed skim milk gained more weight than those supplied only with grain and water.[25] What chemurgists saw as a wasted byproduct, farmers viewed as a valued stock food.

A second challenge for would-be casein manufacturers was competition from casein abroad. Methods to make casein were known by the early twentieth century, but few dairy plants invested in the process because of import competition. Until 1929, the United States imported more casein than it produced, except for the years 1919 and 1924.[26] The competition between imported and domestically produced casein kept prices low and reduced the quality of the domestically produced product.[27] After 1929, increased tariff duties on casein reduced imports and encouraged domestic production. Whereas in 1922 twice as much casein came from abroad than was domestically produced, by 1932 only 6 percent of it was imported. The number of casein factories in the United States increased from 136 in 1921 to 624 by 1937.[28]

The largest and earliest industrial market for casein was for paper. Publishers needed smooth-surfaced paper to print lithographs, magazines, and color advertisements. Casein bound mineral matter to paper, giving it a glossy finish and enabling it to absorb colored ink. But the milk protein could putrefy on the paper's surface—making both the paper and the factory that produced it emit a foul smell. In the late 1920s, chemists and manufacturers perfected the paper-making process, but the natural properties of casein remained. As one chemist admitted in 1927, "It is true that there is still a slight odor in paper coated with casein, but it is due to the casein itself and is not caused in any way by spoilage. Occasionally an individual is found to whom this odor is objectionable but to the greater majority of persons it is hardly noticeable."[29] What better way to bring farmers into the future than to make skim milk into the raw material for advertisers, the so-called apostles of modernity?[30]

Casein also was an ingredient in paint. As in papermaking, casein's adhesive properties made it suitable for binding color pigments to hard surfaces. Casein paint covered surfaces so quickly that it took few coats to apply. Compared to oil-based and other synthetic products, casein paints lacked strong toxic odors. The superior reflectivity of casein paint made it a favorite for brightening factory interiors, especially as manufacturers installed electric lights. At the 1939 New York City World's Fair, the bright paint covering the walls of the exhibit buildings came from casein.[31] Reformulated, casein's innate natural qualities–its tacky texture, subtle odor, and reflectivity—came to symbolize technological progress.[32]

Casein could also become the most synthetic of all substances: plastic. Mixed with dyes and fillers and then soaked in formaldehyde, hard casein plastic could be used for buttons, pen casings, combs, beads, umbrella handles, or even knitting needles and dominoes. Casein plastics were especially desirable to button manufacturers because they could be tinted a wide variety of colors. The buttons had a shiny surface that enhanced garments' flashy appearance. Casein was not the only surplus farm product that could be made into plastic, though, and chemurgists lauded plastics made from wood fibers and oathulls as much as those constructed from milk proteins. By the 1940s, newer plastics challenged casein plastics in the marketplace.[33]

The most dramatic transformation of casein was its reincarnation as an object of high fashion. In 1936, Italian chemists developed a process to make a wool-like fabric from casein, and soon an Italian textile firm began to produce thousands of pounds a day of the product, called "lanital."[34] By 1938, two USDA Bureau of Dairy Industry scientists applied for patents for its manufacture in the United States.[35] The National Dairy Products Company began commercial production of casein wool bearing the tradename "aralac" in 1941.[36] Casein fibers soon replaced fur in hats and were used to upholster car seats.[37] In 1939 and 1941, casein fabrics were at the center of fashion reviews in New York City.[38]

If any casein product symbolized chemurgists' modernist vision, so-called milk wool was it. In casein fibers, scientific innovation seemed to trump natural limits; bizarrely, cows could produce wool![39] Food consumers might have been ambivalent about artificial flavors in their foods, but in fashion, artifice signified novelty and originality. Synthetic fibers could achieve drape and colors that natural cloth and dyes could not. Big ideas about casein spun by the press, however, loomed larger than actual production of casein fibers. Manufacturers never devoted more than 4 or 5 percent of casein to fabric production. By 1948, even National Dairy Products Corporation, the first company to make casein wool in the United States, discontinued making casein fiber.[40]

Casein wool's lackluster success was not unlike that of the casein industry at large. Making nature the foundation for industry proved more difficult than enthusiastic chemurgy promoters had supposed. Even before casein was extracted for industrial production, nature held some sway over how the production process took place. The seasonality of skim milk production, for instance, drove the rhythms of casein manufacture. If factories purchased enough equipment to extract casein from milk in the flush season of summer, it would stand idle in other months. The protein's natural characteristics posed challenges to its use as an industrial product. For instance, casein plastic buttons warped in humid climates. Finally, casein manufacture could not resolve the sanitary problems dairy byproducts posed. Once casein had been extracted from skim milk, over 90 percent of the milk's volume remained as whey.[41] Because the casein extraction process frequently used sulfuric acid, much of the whey byproduct of casein manufacture could not be safely fed to livestock. Hence, rather than eliminate creamery waste, casein production replaced one dairy byproduct with another.

The most pressing problem to casein production was that optimism about transforming skim milk into industrial products could not overcome notions of milk as a healthy food. In theory, using milk to feed industrial processing *and* human bodies promised a more efficient and profitable dairy economy, but in practice, the uses of milk both for industry and for food were incompatible. Casein manufacture created objectionable odors, and most dairy plant owners felt that whatever the economic benefits of casein production, they could not accept a process that might taint the purity of the milk, cream, butter, or cheese they produced.[42] Even as casein production increased in the 1930s, it still constituted only a fraction of the market for skim milk.[43] By the mid-1930s and early 1940s, proponents of chemurgy found themselves competing with public health officials, physicians, and nutritionists who lauded skim milk as a human food, rather than as a fuel for industrial production.[44]

Even as other uses for skim milk prevailed over industrial casein production, the underlying ideas that chemurgy proponents advocated were not lost on the dairy industry in the 1940s. One element of chemurgical thinking was its optimism about making agricultural residues into marketable goods. Wartime conditions intensified concern about a wasted surplus, and such concerns drove the expansion of skim milk production. Further, chemurgy called attention to the promise of organic chemistry's transformative powers. Though chemists of the 1940s and 1950s focused more on casein's use for food than its industrial uses, they were ceaseless in efforts to incorporate whey, casein, and milk sugar into nondairy foods and pharmaceuticals.[45]

Skim Milk Goes to War

While the vision of agrarian modernity promoted by chemists and industrialists for casein was never fully actualized, World War II successfully reincarnated skim milk from a locally utilized byproduct into one with national and international appeal. Although dairy manufacturers knew how to produce skim milk powder at the turn of the century, it was not until World War II that changing cultural mores and expanded government funding for powder manufacture made skim milk a valued human food. Wartime rationale and military-driven contracts for skim milk boosted production and fostered its development. After the war, though, the product faced many of the same challenges as it had previously. Whether skim milk was "trash" or "treasure," and for whom, remained an open question well into the 1950s.

One of the biggest hurdles that skim milk faced in becoming transformed into food for humans was the stigma it carried in the minds of consumers. From the early twentieth century, most Americans considered skim milk a loathsome stepsister of the dairy industry's heralded foods: butter and cream. Skim milk resembled the watered-down milk sold by unscrupulous milk dealers. It tasted thin and lacked the consistency to thicken coffee or be whipped into cream. Advertisers claimed skim milk was less filling, less sustaining, and less whole-some than creamy whole milk.[46] Its use as animal feed limited the material supply of the product available and also affected what consumers thought of it. Convincing people that a substance commonly used as hog slop was healthy and palatable posed a serious challenge.

Though the poor reputation of skim milk caused difficulties, technological developments in milk drying and nutritional research in the 1920s and 1930s set the stage for its transformation. First, creamery operators found that drying skim milk provided a simple way to turn a dairy byproduct into a marketable product in its own right. Dairy companies sold dried milk powder in the early twentieth century but produced more of it in the late 1920s and 1930s as farm families shifted to delivering whole milk to creameries. The reduced bulk of dried skim milk made it easier to transport. Moreover, since dried milk did not have to be refrigerated, it could withstand tropical heat. By 1925, enough dairy manufacturers were interested in dried skim milk that they formed the American Dry Milk Institute to carry out research and promote their product.[47] Aside from its use as animal feed, skim milk powder was used by bakers, confectioners, and ice-cream makers. Skim milk powder had become a product, not a mere byproduct of creamery manufacture. Creameries, previously butter factories, were rechristened as butter-powder plants.[48]

Changing nutritional policy and government distribution of skim milk laid the groundwork for introducing skim milk as a human food. In the 1910s, nutritionists focused so squarely on butterfat's unique status as a carrier of vitamin A that skim milk appeared to be a food lacking in nutritive value. But by the 1930s, nutritionists emphasized that skim milk had positive food values. It provided protein, calcium, phosphorous, and vitamin G (riboflavin), a pellagra preventative. Its milk sugars regulated intestinal health.[49] Promoters of dry skim milk encouraged its incorporation into bread and baked goods, cottage cheese, and ice cream. They also endorsed the use of dry skim milk powder for relief purposes. In the wake of a drought in 1929, the USDA's Bureau of Dairy Industry distributed dry skim milk powder to southern grocery stores and encouraged plantation owners to blend it with cornmeal to stave off malnutrition.[50] When the 1934 Jones-Connally Act earmarked funds for the distribution of dairy products for relief purposes, more and more Americans became familiar with dry milk powder.[51] And yet skim milk's designation as a food for the poor still made it seem less desirable than whole milk and cream. Even those relief clients who used dry milk tended to abandon the product once their economic situation improved, for the item was so strongly associated with being on relief.[52]

Though increasingly aware of the nutritional benefits for humans of their product, dry milk manufacturers in the 1930s still found the animal feed market a lucrative one. They especially promoted the use of dry skim milk for chicken feed. As chicks and broilers began to be raised in cages, their access to green grass, bugs, and sunshine diminished and thus the protein and vitamins that the outdoors provided. Skim milk manufacturers argued their product would nutritionally supplement chicken feed, so that chickens would grow quickly, hatch effectively, and remain free from disease.[53] Thus, the animal feed market for skim milk powder grew in the 1930s.

It took a war to turn skim milk from hog slop or chicken feed to a valued human food. Skim milk entered World War II before American soldiers did. The U.S. Secretary of Agriculture asked for expanded production from dairy farmers in July 1941, months before the attack on Pearl Harbor. Dairy products quickly became essential to the lend-lease and war relief programs. Whereas evaporated milk was the preferred relief food during and after World War I, dried milk powder's transportability and long shelf life gave it the favored spot in the lend-lease formulary. By 1941, the federal government asked for two hundred million pounds of dry skim milk powder for America's allies. Dried skim milk manufacturers struggled to keep pace with the unprecedented demand.[54]

Meanwhile, nutritionists looked to skim milk as a valuable food for civilians. One important advisory council to the War Food Administration was the National Academy of Sciences' National Research Council Food and Nutrition Board.[55] In an August 1942 report titled "The Nation's Protein Supply," the board trumpeted skim milk as an alternative protein source and recommended its more widespread consumption. "The skim milk supply," the report concluded, "represents an extraordinarily abundant source of high quality protein, which at the present time is largely being squandered."[56] Nutritionists' focus on alternative protein sources was not a new development of World War II. The Food Administration during World War I, for instance, urged housewives to institute "meatless" days.[57] But the focus on dry skim milk was new to the World War II food-planning effort.

Besides civilian and lend-lease demands, military rations incorporated skim milk powder. Powdered skim milk was especially important on naval ships in the South Pacific. Military cooks used a special piece of equipment—one that United Dairy Equipment called the "mechanical cow"—to emulsify and blend skim milk powder, butter, and water into milk and cream. By 1943, according to a company trade catalog, "Hundreds of thousands of Uncle Sam's fighting men and Merchant Marines are receiving their daily rations of pure, delicious milk, cream, and ice cream, thanks to the dependable services of the Mechanical Cow."[58] Powdered skim milk reached all branches of the service. In time, the U.S. Army's Quartermaster Corps required skim powder to be blended into bread dough, pressed into cracker wafers, and mixed into the candies for the K ration.[59]

Producers of dried skim milk powder proudly noted their product's role in the war effort. The 1942 annual report for the Consolidated Badger Cooperative, for instance, remarked that "Badger employee's [sic] and Badger farmer's sons, from Australia to North Africa, have had Badger milk on army camp tables."[60] A Rochester, Minnesota, solider fighting in New Guinea sent a letter printed on the back of a dried milk powder label to the Rochester Cooperative Dairy that produced it. He wrote, "Here coconut milk is the only fresh milk we have to enjoy so you can imagine how very much we rely on dry milk to see us through. None of us can have any complaints with workers like you behind us."[61] Knowing that their product was reaching the front lines heartened dairy farm families whose relatives and friends were serving as soldiers.

Whether directed to military, lend-lease, or civilian markets, the newfound position of skim milk powder during World War II also derived from a renewed distaste for waste. In a time of war, the careless disposal of dairy by-products was no longer just a sanitary nuisance; it appeared to undermine the war effort. In a November 1942 letter, John Summe, a dairy plant owner in Covington, Kentucky, reported to the Secretary of Agriculture that 2,000 to 4,000 gallons of skim milk were wasted daily in northern Kentucky for lack of drying facilities. To Summe, more was at stake than sanitation. As he warned, "A gallon of skim milk in the sewer is food for the Axis."[62] If citizens could carefully collect scrap metal for the war effort and forgo purchasing new stockings, shoes, and even steak—then surely the dairy industry should be able to make use of the thousands of gallons of skim milk and whey pouring from dairy plants.

Even with the labor and equipment shortages posed by wartime, production of skim milk powder soared to meet demands for lend-lease, military, and civilian uses. By 1945, the nation's nonfat dried milk powder output reached 642,546,000 pounds, a 38 percent increase over prewar levels.[63] More skim milk than ever was channeled to human food. Whereas less than half of the nation's nonfat milk solids was used as food in 1939, two-thirds was used in this manner by 1944.[64] From farm to factory, the dairy industry answered the call to win the war with food.

But by the mid-1940s, manufacturers began to wonder what would happen to the nonfat milk powder market after the war. Dried milk manufacturers were anxious about intensified competition. To obtain enough supply to fulfill military and lend-lease contracts, wartime salesmen of skim milk powder sometimes had to suggest that their long-loyal customers—poultry raisers, bakers, and ice-cream makers—turn to alternative products like soy flour, cornstarch, and other milk substitutes. Moreover, farm families found substitutes for dried skim milk and whey for animal feed during the war years. Thus, skim milk manufacturers worried that once familiar with other goods, customers would continue to use these alternatives in the postwar era.[65]

The slow development of a home market also troubled dried milk salesmen. It was one feat to get dried milk incorporated into army rations and used for relief purposes, but it was another to convince civilians to use the product. In 1941, the National Research Council's Committee on Food Habits, chaired by Margaret Mead, found that the name "skim milk" carried such negative connotations that few people could be persuaded to use it.[66] Dried milk industry representatives agreed that the name was partly to blame for widespread consumer resistance. To promote one's product as "skim" milk, dried milk manufacturers believed, was to backhandedly insult one's own wares. "We can not continue to say of this large tonnage, 'Here is this refuse; eat it; it is good for you,' and expect to get the producer's total milk into general acceptance."[67] Members of the dairy industry began a campaign to change the name of dried skim milk, and in 1942 Congressman Wright Patman introduced a bill to call the food "nonfat dry milk solids." After congressional hearings in 1942 and 1943, the bill passed and was signed into law on March 2, 1944.[68] Officially, "dry skim milk" now referred only to the product sold for animal feed, while "nonfat dry milk solids" referred to the product used for human food. Dried milk manufacturers trumpeted the law's passage, but the name "skim milk," with all of its negative associations, stuck in the vocabulary of most Americans.

Hence, as the war drew to a close, marketing, not production concerns, loomed large in the minds of dried milk companies. Manufacturers recognized that producing dried nonfat milk at record levels would not ensure the industry's success. Without a market for the product, dairy farmers would once again be faced with the surpluses that had plagued them in the 1930s. By February 1947, dry milk surpluses were mounting, even before the flush summer season of milk production. Nonfat milk producers looked to three strategies for readjustment: using nonfat milk for relief purposes, incorporating milk solids into processed foods, and turning skim milk into a diet food. Though each of these strategies drew from historical precedents and skim milk's healthful properties, each also sought to meet the needs and demands of the postwar context.

The nonfat milk manufacturers' first instinct was to lobby federal agencies to distribute dried milk as a relief food. Dried milk had served this purpose before the war, and by 1945, nutritionists redoubled their efforts to improve human nutrition. Extension agents in North Carolina, South Carolina, and Arkansas shared recipes and schooled women on how to mix nonfat dried milk with water. After the passage of the National School Lunch act, nonfat milk was incorporated into schoolchildren's lunchtime meals. In late 1946 and 1947, New Mexico and Alabama schoolchildren in towns too small and remote for fresh milk delivery and refrigeration drank nonfat skim milk as part of their school lunches.[69] By 1949, the USDA directly purchased 9.2 million pounds of dry milk solids for this purpose.[70] The use of dried milk in school lunches expanded earlier programs of relief feeding that had their origins in the Great Depression and set the stage for later state efforts at improving nutrition, such as the War on Poverty campaigns.[71]

International opportunities for relief feeding also drew the attention of dairy farm families and creamery operators. Reports of a war-torn Europe piqued the interest of dry milk manufacturers, for their product—with its high nutritive content, low cost, and easy transportability—was especially well suited to aid in worldwide relief efforts. Dairy manufacturers were fully convinced that if dried milk had helped win the war for the Allies, it could also become a tool in combating communism. In a 1947 letter to his congressman, A. G. Schultz, of A. G. Cooperative Creamery, positioned nonfat milk into new American policies of containment, saying, "If we feed Europe we will not have to worry about Russia."[72] Federal officials in the Department of Agriculture agreed that using the milk surplus for relief made political and economic sense. In 1948, the Department of Agriculture purchased 138 million pounds of nonfat dry milk solids for the United Nations Children's Fund, for army feeding in occupied areas, and other foreign relief uses.[73]

The use of nonfat milk for relief in the postwar era depended on continued congressional funding. Dairy industry officials had put confidence in the school lunch program, but by 1952, some states and schools were suggesting that milk be an optional, rather than mandatory, part of meals served in schools.[74] International relief programs also faced opposition. During the U.S. occupation, Japanese citizens fell ill with dysentery after consuming milk solids that had been blended with unsanitary water, foreshadowing international misfortune with baby formula in the 1970s.[75] By 1951, even as troops departed for Korea, government purchases of nonfat milk powder dropped to 87 million pounds from the 138 million pounds purchased just three years earlier.[76] If dairy manufacturers and farmers were going to find a steady market for an ever-increasing amount of skim milk powder, they had to have a more reliable market than the state and relief agencies alone.

A second approach that dried skim milk manufacturers took in the postwar era was finding ways to incorporate their product into the burgeoning postwar manufactured food market, especially by increasing home consumption and incorporating dried milk into processed foods. Improving skim milk powder's packaging was key to efforts to increase home use of dried milk. Before and during the war, milk powder went to market in large drums or barrels. Bakeries, the largest outlet for dried skim milk, bought the powder in bulky sacks. But bulk packages were burdensome and ill-suited to the home market. Dried milk absorbed the air's humidity in the large boxes in which it was sold. To make nonfat milk solids a consumer good, the product would have to be attractive to the postwar shopper.

Flashy packaging was especially important as food purchasers began shopping in supermarkets where impulse buys were more common.[77] In 1953, the Consolidated Badger Cooperative introduced a new container for its nonfat skim milk powder and sold the food with a special mixing pitcher in which purchasers could reconstitute it.[78] Attractive packaging, combined with high fluid milk prices, boosted nonfat dry milk consumption in the early 1950s. Whereas consumers bought only 2.4 million pounds of nonfat dry milk solids packaged for home use in 1948, by 1951, they purchased 60 million pounds for this purpose.[79] Nonfat dried milk was especially promoted for use in emergency situations. By the 1950s, the food was a regularly recommended staple for equipping bomb shelters in the event of nuclear attack.[80] Though dried milk existed before the war, the manufacturers remade the product to fit the 1950s context, dressing it up for supermarket shelves and billing it as a sustaining food in case of nuclear holocaust.

While some consumers wittingly purchased skim milk powder, others purchased nonfat milk solids as an ingredient in the dried, canned, and frozen foods that became popular in the postwar era. Like chemurgists who had enthusiastically pledged to turn farm surpluses into industrial products, food scientists in the late 1940s and 1950s were optimistic about the power of science to turn surplus agricultural goods into profitmakers. Historian Harvey Levenstein calls the period the "Golden Age for American food chemistry."[81] Touring a suburban home in 1952, one could open the cupboard and encounter nonfat dried milk as an ingredient in prepared mixes for biscuits, cakes, and pie crusts. On a Saturday evening, families might step out onto their patio or porch and enjoy hot dogs that incorporated dried milk as a filler.[82] When teenagers went to the drive-in, they often consumed nonfat milk solids in their soft-serve cones.[83]

Figure 3.3 Marketing powdered skim milk for mass consumption required repackaging the product for retail sale. Consolidated Badger Cooperative Creamery Records, Wisconsin Historical Society, Green Bay Area Research Center.

Nonfat milk was not the only dairy byproduct to see new life through the miracles of food chemistry and food processing. Whey, lactose, and casein also found their way into processed food products and pharmaceuticals. Lactose (milk sugar) became the preferred steeping medium for the mold used in making penicillin. Bureau of Dairy Industry officials incorporated whey into baked goods, candies, and sherbets.[84] Casein, too, came onto the market in new guises. According to a 1954 advertisement printed in the *National Provisioner*, the use of sodium caseinate (milk protein) in sausage would eliminate fat pockets, reduce smokehouse shrinkage, and increase its shelf life. Casein was also hydrolyzed to take on a "brothlike flavor" making dried soups taste like chicken or beef.[85] Strikingly, a food once billed as "nature's perfect food" was undergoing radical transformations to taste like something else. Once milk became a sum of its nutrient parts, these parts could be transformed and substituted for other proteins. The mutability of milk carried great potential for its entry into new markets, but the miracles of food science could also undermine existing markets for dairy foods by making other foods stand in for milk fats, proteins, vitamins, and minerals.

Perhaps it was this vulnerability of dairy foods to substitutes that led to the third strategy milk companies used to promote skim milk in the postwar era: selling it on its own merits. As prices for whole milk increased in the late 1940s, milk dealers in the fluid milk market, not just dried milk dealers, turned to skim milk as a promising product in its own right. Though many consumers were skeptical about the value of skim milk, dairy companies enticed them with promises that drinking skim milk would help them lose weight.[86] Milk dealers secured the backing of physicians. As had been the case for certified and pasteurized milk in the Progressive Era, the recommendations of physicians gave skim milk newfound legitimacy. Although physicians had long suggested nonfat milk to patients who had difficulty digesting fats or were elderly, weight-conscious consumers became the largest sector of the skim milk market in the 1950s.[87] Emphasizing skim milk's role in promoting slenderness transformed skim milk's reputation as a low-cost relief food to one that high-income dieters would embrace.

The desire to diet drew largely from consequences of postwar abundance. Equipped with deep-freezers and family-size refrigerators, Americans had access to a greater quantity of food at home than ever before. As Americans settled into the sedentary culture of drive-ins and television entertainment, many gained weight. Skim milk held a prominent place on the dieters' menu. Gaylord Hauser's *Live Younger, Live Longer*, which topped the best-sellers' list in 1951, called skim milk a "wonder food."[88] Diets printed in women's magazines and even in *Hoard's Dairyman* recommended the food. A Sealtest skim milk offer gave consumers the chance to get a new bathroom scale with a proof-of-purchase tag.[89] As concerns about cholesterol and heart disease intensified, the idea of skim milk as a healthy food would only become more entrenched.

One of the most striking features of dairy companies' push on skim milk as a diet food was the feminization of the product. In the past, milk marketers had entreated women as mothers to buy milk to ensure children's health and safety, but dealers seeking skim milk buyers courted women as an adult market.[90] Skim milk advertisements emphasized not innocence and wholesomeness, but svelte sexuality. A 1950 advertisement printed in *Milk Dealer*, for instance, encouraged milk salesmen to pursue the low-fat milk market by picturing a bikini-clad beauty and the words "She Can Be Yours."[91] Advertisements for Sealtest skim milk suggested that by sipping skim milk, an overweight woman could lose pounds, regain a youthful complexion, and win back her man.[92] Selling skim milk as a

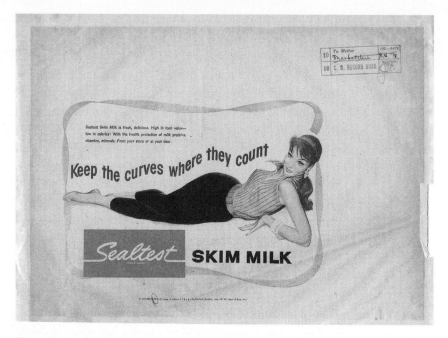

Figure 3.4 By the 1950s, the transformation of skim milk from hog slop to sexy diet food was complete. N. W. Ayer Advertising Agency Records, Archives Center, National Museum of American History, Smithsonian Institution.

potion of slenderness, and thereby romance, certainly gave it more prestige than it carried as a mere byproduct of buttermaking or as hog slop. The innovation of skim milk in the postwar era was not so much in a transformation of the product itself—as chemurgists had lauded—but in its perception. By creating unique cultural ideologies for dairy foods, milk dealers hoped to create a lucrative, segmented market in which dairy consumers might purchase skim milk and cottage cheese for adults, whole milk and chocolate-flavored milk for children, and dried-milk-infused mixes, hot dogs, and baked goods for the whole family. Such cultural transformations had real implications on the farm. By 1960, farm families sold 88 percent of all the skim milk they produced, compared to 50 percent in 1940.[93]

While the dairy industry celebrated its successful re-creation of skim milk into a human food, each of the strategies on which dairy manufacturers relied to sell skim milk carried risks. To maintain their markets, milk manufacturers needed the continued support of the state to channel their product abroad. Further, while skim milk benefited from developments in food chemistry, its manufacturers needed protection from food substitutes as chemists applied the same techniques to other byproducts. Finally, dairy manufacturers could only go so far in emphasizing skim milk's slimming properties, for highlighting the food's low-fat content might only add to public concerns about the detrimental effects of butterfat. Since most skim milk was a butter factory byproduct, the fate of skim milk was tied up with that of butter. Each of the marketing strategies to which skim milk was attached, then, showed signs of vulnerability, even as skim milk entered more pantries than ever before.

Reconceiving Nature: Artificial Insemination in Dairy Herds

If manufacturers viewed dairy wastes as peripheral products to butter or cheese, dairy farmers thought of male calves and unproductive cows as byproducts—secondary to attaining high milk yields. Since male calves couldn't convert hay and grains into milk, they were the most likely to be defined as waste. Thus, most dairy herds had far more milk cows than bulls. When a valued cow birthed a bull calf, it was often greeted with frustration because the male calf would never become a milk producer. As one Vermont farm manager noted in 1939, "Bessie had another bull which was disappointing. I had hoped for a heifer from her."[94] Bull calves raised by leading dairy breeders might survive to adulthood, but most male calves were butchered while young.[95] Farm account books indicate the calves' fate; Illinois farmer Dale Beall's 1940 record book reveals three checks received for the sale of veal calves, all of them in the spring when most cows had their young. Once calves left the farm, it is difficult to track what became of them. But the checks Beall received from Oscar Meyer & Company for calves sold in 1949 and 1954 provide some indication.[96] After leaving the farm, muscles of most male calves became veal and their hides "calfskin leather." Just as dried skim milk boosted the profits of butter manufacturers, the proceeds from bull calf sales supplemented farmers' earnings from cream and milk.

Male calves were not the only animals that dairy farmers classed as byproducts of farm production. Farmers valued cows that were able to convert a low-cost ration into milk efficiently so that the value of the milk produced exceeded the cost of feeding them. A cow that consistently produced less milk than others drained farm profits, for she continued to require feed and labor without offsetting these costs. Cows that had troubles breeding were also problematic, for failure to carry a calf negated cows' milk-producing potential. Farmers reviewed the balance sheet more strictly in light of high feed costs, seeking to determine which cows' appetites were squandering farm profits. As farm owner Virginia Hollerith asked a farm manager, "Gotten any of the doubtful ones bred? I would like to give them a chance but do not want to keep on feeding the hopeless ones."[97]

Hollerith was frank in her assessment of nonbreeding cows as "hopeless," because the main measure of dairy farm profit was how much income a farm obtained from the sale of milk over the cost of cattle feed. One way to boost profits was to cut feed costs by raising hay and silage on the farm, thereby reducing off-farm purchases. But another important way was to boost the amount of milk cows produced by breeding cows with bulls whose offspring had a record of high production. In the late 1930s and 1940s, developments in cattle breeding—most notably the advent of artificial insemination (AI)—promised to heighten cows' milk-producing capacity to new levels.

Advocates of AI believed they could improve farm prosperity by making dairy breeding more efficient. As sperm from prize-winning studs replaced the bodily functions of bulls, barnyard bulls became byproducts of dairying. AI carried important implications for how farm people understood and interacted with the animals on their farms. In the same way that tuberculin testing made bacteriology as important as clinical observation of animals to gauge animal health, the adoption of artificial breeding encouraged farm families to think of cows in measured scientific and numeric terms and as technologies to be managed, rather than as animal actors capable of their own reproduction.

Decades before farmers adopted AI, dairying experts urged farmers to think of cows and bulls as figures in a ledger, rather than figures of flesh.[98] The invention of the Babcock butterfat tester in 1895 enabled farmers to measure how much milk and butterfat individual cows produced. When faithfully recorded, farmers could use these measurements to identify the highest and lowest-performing cows in their herd. Reformers encouraged farmers to compare the production records of dams and their daughters. If calves produced more milk than their mothers, it suggested that the sire had a beneficial effect on milk production. From the 1910s to the 1930s, these record-keeping and testing tasks became systematized in organizations called "cow testing" or "dairy herd improvement" associations (DHIA). Farm families joining a cow-testing organization paid an annual fee to fund the salary of a full-time employee who traveled to member farms and recorded the feed consumption, milk production, and profit for each cow. Based on these figures, farmers could determine which cows did and did not contribute to farm profits and use the information to cull unprofitable cows and breed for improved efficiency.[99]

However, not many farm families turned to record books to ascertain which bulls and cows were valuable, and which were best classed as waste. Few farmers kept records as faithfully as experts encouraged. In 1936, fewer than 2 percent of the nation's dairy cows were being tested by herd-improvement associations. Further, although ideally, farmers would be able to determine a bull's quality based on its pedigree and the record of production of cows it had sired, bulls with proven records sold at such high prices that hardly any ordinary dairy farmers could afford to use them.[100] The Hollerith family in Virginia was the exception when it purchased a bull in 1932 from Klondike Farm in Elkin, North Carolina. The bull's seller advised the new owner, "I have never seen in any one herd any better type or producing cows than his sisters...and the dam of this young bull is one of the five cows of the breed that has three records that average 700 lbs. fat before she was five years old." To obtain such a promising bull, the Hollerith family spent $330.[101]

Hence, well into the 1930s and 1940s, lacking records, most dairy farmers made breeding choices by finding good-looking animals on nearby, well-kept farms, not by consulting record books. In 1922, Elihu Gifford visited a farm just eighteen miles away to look for a quality bull calf and then purchased it for $20.[102] Gifford read the body of the cow, not the records kept by the cow's owner, to inform his decision. He used the masculine build and body-line of a male calf, in lieu of pedigree, to ascertain the bull's future as a herd sire. The values an animal carried were so embodied in its physical form that sometimes farmers dispensed not just with records, but with cash as a means of exchange in bull purchases. Such was the case when the Sherwood family, who operated a farm near Windsor, New York, obtained a Guernsey bull in 1932 by trading a heifer.[103]

While some farmers purchased bulls, others relied on the bulls of neighbors to breed their cows. Wisconsin farmer Harvey Dueholm, for instance, remembers that his father led cows to a nearby farm to breed with the neighbor's registered Guernsey bull. The Dueholms paid two to five dollars every time they had a cow serviced, which likely offset the high cost the neighbor paid for the animal. Mr. Dueholm's memories of the breeding encounters are sharp, despite his father's efforts to shield his son from the details. "They didn't want us to see it happening...they always took the cow back behind the barn and I had to stand on the milk stool and look out the window!"[104] For the farmer walking a balky cow across the pasture to a neighbor's ready bull, animal breeding was hardly an abstraction. The circumstances of the act of breeding left much control to the animals, even as it was commoditized and guided by farmers' choices to seek out a registered animal. Clearly, the young Dueholm understood the act he witnessed surreptitiously to be a fleshy bodily encounter, not simply a theoretical pairing. Lacking records, farm people's role as breeders of livestock demanded that they be attentive to the intimate longings of the animals in their care, and skilled judges of their bodily capacities.

Indeed, many farm people understood all too well and too directly the power inherent in their animals' bodies. Caring for bulls was demanding and dangerous. Bulls could act unpredictably and wildly. The same bull that appeared calm and placid could unexpectedly snap in a fit of rage, goring or trampling the humans who tended him. The diary of Boyd Sherwood, who lived on a farm in southern New York, provides evidence of one such encounter. On October 5, 1940, Sherwood recorded that "Leslie Turrell was attacked by his big bull and they don't expect him to live." Indeed, Mr. Turrell died a night after the attack.[105] To keep a bull safely, farmers built high-fenced bull pens and set aside at least an acre of farmyard for the bull to graze. For farmers with small herds, especially, the expense and trouble of keeping a bull were steep. Even larger farms found keeping bulls a challenge; dairy breeder Virginia Hollerith consistently clipped articles about how to construct safe and durable housing for bulls.[106]

Attack wasn't the only risk that bulls posed. Animals used for natural service stood a great risk of contracting reproductive diseases because of their frequent sexual encounters. Once infected, bulls could easily become disease vectors, spreading granular vaginitis, vibrio fetus, trichomoniasis, or Bang's disease from herd to herd. Venereal diseases of livestock often caused sterility and triggered abortions. A sick cow would reduce a herd's profitability in milk production, increase a farm's veterinary costs, and damage a farm's reputation for livestock breeding.[107]

It was within this context—in which most dairy families made breeding decisions without adequate information and experienced great risks of disease and bull attacks—that dairy researchers introduced the practice of AI in 1937. In the new procedure, rather than merely choose specific bulls to mate with selected cows, a dairy farmer could hire a veterinarian or inseminator to collect sperm from a specified bull and inject it into the selected cow's reproductive tract. As AI studs, bulls would move into pens or stalls designed to facilitate regular collection of semen.[108] About once a week, the inseminator or veterinarian would lead the bull to a restrained cow or a "dummy cow" made of steel and padded canvas. Just as the bull was about to mount for mating, the veterinarian would redirect the bull's penis into a long rubber-lined tube, smeared with petroleum jelly. A layer of warm water between the shell and the lining of the tube simulated the pressure and warmth of a cow. After the bull climaxed, his semen would be poured through a funnel and into a test tube.[109] AI still relied on a bull's hormonal urges for its success. Bulls had to be sexually stimulated to discharge the semen that inseminators needed. But by reconfiguring the act of reproduction from an animalistic act of copulation to a human-controlled exchange of fluid from one body to another, dairy breeders believed they could overcome the uncertainty and natural variability that had limited existing herd-improvement efforts.

Like chemurgy, AI was a technology promoted as a way to reduce waste. In nature, bulls produced millions of sperm in each ejaculation, only one or two of which would become a calf. AI took thousands of "wasted" sperm and turned them into profitable calves. Bred naturally, even the most potent bull could only produce thirty or fifty calves a year. But aided by human hands, a bull's semen could impregnate thousands of cows. In the words of a genetics investigator for the Bureau of Animal Industry, AI could correct "profligacy of nature."[110]

AI also reduced the necessary time to prove a bull's worthiness. Semen from young bulls could be collected and used to fertilize cows before bulls were of sufficient size to mount a large female cow. The milking records of offspring from cows so fertilized could be assessed more quickly to determine whether the bull had a beneficial influence on milk production.[111] Even more important, once a bull obtained a solid record of improving milk production, dairy breeders could extend the reach of the best bull's breeding beyond natural limits. Even bulls whose ability to mount cows was hampered by injury or age could be used for AI.[112] A greater number of farm herds could utilize prize-winning semen than could be serviced naturally by an elite-class bull.

For farm families dependent on a milk check, there was much to like about the new practice.[113] Farm people welcomed lower feed bills and appreciated the opportunity to rid the farm of the danger a reckless bull posed. At 60 percent, the conception rates for AI rivaled those of natural service, and due to careful selection of sire semen, AI offspring often bested the milk production of cows bred naturally. Lower costs, fewer risks, and better production rates—these were gratifying developments for farm people who had weathered the volatile economic conditions of the 1930s and worried about a postwar economic slump in the wake of World War II. By 1946, 579,477 American cows had been bred artificially. Three years later, in 1949, two and a half million dairy cows, 10 percent of the nation's total, were enrolled in AI associations. By 1958, the number of cows and heifers bred artificially reached seven million.[114]

The records of the Mallary farm, located on the Connecticut River near Fairlee, Vermont, provide a close look at the changes arising with the shift to artificial breeding in the 1940s and 1950s. In 1935, R. Dewitt and Gertrude Mallary purchased forty-two acres previously utilized for growing strawberries and celery and transformed it into a leading Holstein-Friesian dairy in the northeast. The Mallary farm was unlike most dairies. Farming was not the Mallary family's sole source of income; Dewitt Mallary worked as a corporate lawyer in Springfield, Massachusetts, and hired laborers to perform much of the farm work. The family purchased nearly all its machinery new and was able to construct or remodel many farm buildings—including the dairy barn—in the years immediately following the farm's purchase. It stocked the farm with purebred Holstein cattle obtained from a dairy breeder in Connecticut. Despite these differences, the records of the Mallary farm lend insights into how farm families adopted AI in its earliest years.[115]

The Mallary family and others who used it needed more than a potent bull and ready cow. They also had to become further enmeshed in networks of technology and expertise. Farmers had to have access to telephones and good roads to take advantage of AI.[116] Phones helped farmers quickly inform the veterinarian of a cow in heat and roads enabled the inseminator to reach the cow at the designated time. AI also required regular visits from technicians to "service" cows and advise farmers about how to overcome cows' sterility problems or care for them during pregnancy.[117] By extension, AI relied on educational institutions, such as agricultural colleges and veterinary schools, to train inseminators. Mr. and Mrs. Mallary regularly sought advice from Dr. Walter W. Williams, a Cornell-trained veterinarian who wrote the 1929 book *Sex Hygiene and Reproduction of Cattle*.[118] Those who were geographically isolated in non-dairying regions, lacked telephones or good roads, or had no access to experts had few chances to take advantage of the new practice. Like many other farm improvements, the practice of AI relied on a host of technological transformations.[119]

Not all farmers welcomed the expanded role for experts in farm management that AI necessitated. Frank Hill, who tended dairy cattle near Knoxville, Tennessee, was a charter member of the region's artificial breeders association, and he liked that the bulls available through AI were superior to those in his region. Even so, he admitted that he found using a neighbor's bull more convenient than artificial breeding.[120] Relying on the inseminator to service cattle meant confining the cow to the barn and waiting for the inseminator to arrive, a considerable inconvenience during cropping season.[121] Some farmers who had long resented the intrusion of "white-collared" university experts resisted allowing agricultural professionals to gain any more control over farm production. In time, some farm people adapted to AI by gaining the veterinary skills and taking on the managerial responsibilities once performed by inseminators. Frank Hill's son, for instance, trained so he could do the work of the inseminator for his herd. Initially, practicing AI meant welcoming not just the ideas of industrialization but its most enthusiastic agents onto one's property.

On the Mallary farm and others, putting bovine reproduction into human hands introduced a host of new questions whose answers had previously been left to nature. How long could semen be stored? How much semen was necessary for a cow to be impregnated? Where, exactly, should semen be deposited for the best results? For the managers of the Mallary farm, microscopic examinations of semen, rather than assessment of a bull's body type, became increasingly important to the work of dairy breeding. Farms that bought semen also utilized microscopic analysis to determine sperm's potency and vitality. As early as 1940, just three years after AI was introduced in the United States, Mr. Mallary purchased a microscope so that the farm's manager could examine semen samples before artificially inseminating cattle. In 1954, a Massachusetts client had a cow come into heat nearly two weeks after receiving semen from the Mallary farm. The client explained, "I had the inseminator examine it under his microscope. He found that 5 percent of the semen were still alive and very active so he used it to inseminate the cow. We were all astonished that semen 11 days old was in such good condition."[122]

The very notion that sperm had to be examined for freshness marked the change that had taken place in the ways that farm people knew and understood the animals on their farms. Once, farm people had witnessed bulls mounting cows in a pasture or the barnyard. They hadn't needed to ask whether or not the bull's sperm was "fresh." Now, they stood by as trained inseminators came to the farm with specially designed flasks and funnels, investigated semen, and charted cows' estrous periods. The outcome of dairy breeding was still new calves, but the act of a bull copulating with a cow had been reduced into a series of steps to be directed, analyzed, and repeated on farms throughout the country. Even an unruly bull's most intimate desire was now handled by scientists and bureaucrats.[123]

Expertise and technological developments could not crack all the mysteries of bovine reproduction. To be successful, cattle breeders needed to carry out a variety of tasks that bulls' biological urges had once directed. For instance, inseminators had to learn how to detect whether a cow was in estrous to ensure that semen would be inserted at just the right moment. But defining "the right moment" for conception was difficult.[124] Estral cycles varied among animals and some cows showed few signs of estrous at all. In that case, nature proved a more reliable guide than farm records. One AI expert recommended using yearling steers to detect cows in estrous, for "they retain their sexual desire for several months following castration."[125] Though displaced from directing the task of reproduction, bulls still had specialized knowledge that humans found difficult to replicate. Bulls' organismal qualities made them irreplaceable to the production process.[126]

Further, though bulls had been shaped and selected to serve human purposes, they were still creatures, prone to disease and parasites like their kin in more "natural" settings.[127] Seasonal variations in fertility stifled human efforts to turn cattle into breeding machines. Scientists attributed seasonal decline in cattle's fertility level to variations in sunlight and its effects on cows' and bulls' pituitary glands.[128] The predictability, reliability, and efficiency that AI promised had limits. Bringing mass production to dairy breeding meant adapting cows and bulls to the industrial scale. It also meant adapting industrial production to the processes of nature.[129]

It is hard to assess what the adoption of AI meant for the bulls and cows on which the practice relied. Though dependent on bulls' vigor and cows' ability to carry a calf, AI minimized animals' direction over reproduction and the centrality of animals' bodies themselves. Bulls used for AI performed the tasks of reproduction but did not direct the process. On many farms, producers had previously "managed" reproduction by allowing bulls to run with the herds and mate at will.[130] The fertility of bulls and cows peaked in the spring and summer months, just at the time when farm families were busy with other farm tasks and cattle grazed on pastures for much of their ration. Hence, during the spring and summer, bulls on some farms had mated with cows outside of the purview of their owners.[131] Such couplings were not isolated to farms that paid little attention to breeding. In 1942, for instance, the Mallary farm manager reported an instance of a cow that calved as a result of an unrecorded breeding. He wrote, this "must be by Don Inka as he broke out of his pen on several occasions and I know of no other answer."[132] By separating bulls from cows more completely, AI disallowed bulls from determining the circumstances and frequency of mating with other animals.

The discovery of methods to freeze dairy semen in 1950 marked an important breakthrough in AI. Frozen semen could be stored for years and travel long distances, as long as it remained cold. The use of frozen semen simplified the insemination process, enabling farmers to purchase semen in individually measured doses. Though inseminating cows still required farmers to be able to detect estrous and to chart breeding schedules, more farmers could practice insemination of cows themselves.[133] The transportability of frozen semen also enabled AI to surmount the geographical limitations that had previously hampered the development of the AI in nonspecialized dairying districts. By the 1950s and 1960s, the Mallary family obtained semen from Wisconsin, California, Pennsylvania, Indiana, Tennessee, and Ontario to use in dairy breeding.[134]

AI altered the ways in which farm families conceived of their dairy herds and further minimized the importance of keeping bulls on the farm. The widespread use of frozen semen made dairy breeding more efficient and more disembodied. Once a bull's semen had been frozen, the bull's body became almost superfluous. Frozen semen, unlike a bull's natural body, could outlast accident, disease, and age. Farm families devoted funds once used to pay for feeding a bull to semen storage costs. It was not just bulls' bodies that disappeared from the farm. As cows' capacity to produce milk increased, farmers eliminated low-producing cattle. By the 1950s, Americans relied on fewer and fewer cows to produce more and more milk, a trend that would only increase in the postwar period.

Looking more closely at the introduction of AI of dairy cattle helps expose how the industrialization of agriculture changed nature and farmers' relationship to it. Domesticated animals had long borne the marks of human desires. Over centuries, people narrowed the range of natural variability among cattle. Animals with the greatest chances of reproducing were not those that best adapted to their environments, but those who displayed traits that humans deemed useful. AI further constricted animals' freedoms to meet humans' wishes. Bulls' rations were carefully measured, exercise regularly timed, and their "service" attentively assessed. Though the diet and lifestyle as a stud bull probably bested that of bulls on many farms, it did reduce their opportunities for bullish behavior, such as fighting with other bulls or mounting a cow in heat.[135]

Farmers justified and even applauded animals' loss of control over the reproductive process because of the social and economic benefits AI engendered.

Once farmers began using AI, they could sell the bull, use the ration of grain and hay it had eaten to feed milk-producing cows, and escape the danger that had lurked in the bull pen. The reduced risks of disease, attack, and low milk production helped farm people adjust to a world in which cows were fertilized in "unnatural" ways—not by the bull, but by having a human arm reach up a cow's rectum to position the cervix, easing entry of a long glass tube that injected semen into a cow's uterus. Though AI was often promoted as making cows more akin to machines, those who regularly interacted with breeding animals knew that cows and bulls behaved differently than tractors. Even as it made breeding more selective and controlled, AI still relied on cows' and bulls' natural proclivity to produce milk and semen. Cows kicked. Bulls stampeded. Cows' estrous periods were hardly signaled by flashing indicator lights. Many farmers still experienced uncertainty over whether a cow would successfully conceive. Everyday experiences with unpredictable animals simultaneously tempered human hubris and fostered receptivity to technologies like AI.

Even in an age of AI, breeders became attached to individual animals, and their appreciation for bulls extended beyond the modern, industrial criteria praised in breeding records. In 1944, a prized bull on the Mallary farm, Posch, died of heart failure. The Mallarys' reaction betrays intimate ties to the animal, not mere abstraction. As Mr. Mallary wrote, "I miss him very much already, not only because he was a great sire, but because he was a delightful personality. It seems strange not to greet him every time I go into the barn."[136] Just as animals retained an independence and other-ness, no matter how completely humans transformed them to meet human initiative, so also did humans' relationship to animals maintain some semblance of intimacy even after AI was adopted. Indeed, it was precisely because of this tight connection to individual animals that AI appealed to dairy breeders. Through utilizing stored sperm, the "delightful personality" of beloved bulls like Posch could live on as calf progeny.

Environmental historians have often told the story of agricultural industrialization as one in which humans' ties to the environment become increasingly attenuated and abstract. Consumers still relied on nature, but they became distanced from it—physical and conceptually.[137] In many ways, the story of AI parallels this common pattern. With AI, dairy farmers' relationship to bulls was increasingly remote and disembodied. Experts implanted semen from a distant bull that boasted a strong production record. Cows no longer copulated in a nearby field. Less often appreciated by environmental historians, however, is how and why an abstract, alienated relationship to nature was ever alluring. For farm producers, a disembodied nature was one that carried fewer risks of bull attack and lesser likelihood for animals to suffer from reproductive ailments. Further, it was one in which valued animals could live beyond their "natural" limitations—size, injury, death. The case of AI, then, helps reveal why achieving a more distant and remote relationship to nature seemed a positive development, even among farm families who most recognized and valued the importance of animals to human life.

Getting in Touch with a Changed Nature

At the very moment that farm families reorganized their breeding practices and dairy herds to distance themselves from the risks of nature, consumers in the postwar era renewed their quest for direct and authentic relationships to nature through recreation. Record numbers of leisure-seeking canoeists, wildlife enthusiasts, and anglers took to the outdoors in the 1950s, and as they did so they encountered dairy byproducts befouling streams near dairy plants. Whey, especially, continued to be a problem, for while nonfat dried milk found new markets, research and markets for whey proteins lagged behind. When too much waste flowed into the waterways near dairy plants, a heavy black sludge coated them. Aesthetically unpleasant and odiferous, putrefying dairy waste depleted oxygen in streams and diminished fish populations.[138]

Thus, in states like Wisconsin, with many dairy plants and recreation-seekers, outdoors enthusiasts rallied around curbing water pollution. Members of the Izaak Walton League and state legislators mounted a campaign to increase fines for water pollution and require compliance with existing water pollution regulations. Sportsmen and -women argued that the state's waters belonged to all and should not be tainted for private gain. Initially, the Wisconsin Cheese Makers' Association opposed such efforts, arguing that the state should not toughen enforcement until adequate research had been carried out to dispose of waste properly. But by 1950, dairy plant owners began cooperating with the state's committee on water pollution, even inviting a member of the committee to speak at their annual meeting.[139]

Dairy plant owners' acknowledgment of the water pollution problem likely stemmed from two sources: the declining ability for local farmers to address the byproducts problem and the concentration within dairy processing. First, the postwar trend to specialization in farming meant that a long-utilized local method of waste disposal—animal feeding—was no longer sufficient in curbing dairy waste. In the 1920s and 1930s, sanitarians who wanted to reduce the flow of dairy wastes into streams had advised cheesemakers to return whey to farmers for use in animal feeding. But in the 1950s, the number of farms with livestock diminished. Farm families began to reorient their farms to maximize production with less labor. Some farm families ridded their farms of hogs and chickens to focus on dairying, while others specialized exclusively on beef or hogs. Thus, fewer farmers lined up at the dairy plant clamoring for whey or skim milk to feed their nondairy animals. Furthermore, as farm families substituted vitamin-enriched dried whey and dried milk concoctions as calf or chicken feed, they did not utilize the byproducts of dairy manufacturing directly as they once had. Changing farming practices, not just suburban and urban environmentalism, contributed to a greater focus on the dairy waste problems.

Another trend that stimulated attention to dairy waste was a concentration in the dairy manufacturing industry. As rural transportation improved and the processes of dairy manufacture became more complex, many crossroads creameries and cheese plants closed. Farmers' milk traveled farther distances to modernized factories equipped to handle a variety of milk manufacturing processes: cheesemaking, buttermaking, milk bottling, and sometimes producing ice cream or powdered milk. Consolidation in dairy processing increased the volume of waste plants produced and concentrated its effects. Wisconsin's dairy plants produced 35 percent more waste in 1954 than in 1930.[140] Even after new technologies to utilize dairy byproducts were developed, an imbalance between production and consumption of dairy byproducts remained.

In seeking to minimize inefficiencies at every level of the dairy production process, dairy farm families and manufacturers created new kinds of waste. AI made sperm distribution more efficient, but it also required finding new uses for dairy bulls. New calf feeds channeled nutrients more effectively to young calves, but generated a surplus of skim milk. Specialization within agriculture and dairy manufacturing made the use of labor more efficient, but it also broke down the local cycling of byproducts between farm and factory that had once been the solution for minimizing waste, albeit an imperfect one.

Although recreationists and dairy manufacturers both expressed frustration about water tainted with dairy waste, their objections to the problem stemmed from different sources. Recreationists objected to dairy wastes because they undermined the aesthetic benefits they sought from nature; they sought to purify streams so that an afternoon spent canoeing or fishing provided an escape from the detritus of daily life. Dairy manufacturers, on the other hand, worried about the inefficient use of resources that wasted whey or milk represented; they sought to channel waste into more productive (and profitable) pursuits. In this way, the humble and little-known history about dairy waste mirrored earlier debates about the proper role of nature in modern life, including those as pivotal as the pitched battle over the fate of the Hetch-Hetchy Valley outside of San Francisco. Both cases pitted those who valued nature for its aesthetic benefits against those who valued its human utility for the greater good.

But dairy manufacturers could not dismiss recreationists' concerns for aesthetic nature completely, for consumer concerns about unhealthy landscapes and impure air could undermine markets for cheese, butter, and milk. Thus, dairy plant managers sought to harness consumerism to remedy problems with waste. Research in chemurgy as well as new products like nonfat dried milk attracted manufacturers' attention for they seemed to promise to eliminate the aesthetic harms dairy waste posed and also fulfill their mission of utilizing nature efficiently. However, dairy manufacturers' consumer-driven solution was as dependent on the new byproducts' cultural acceptance as their material transformation. Scientists could develop ways to turn whey into a creamy spread, make casein into paint, and transform skim milk into pimento-speckled cottage cheese, but they could not guarantee that people would buy the new products. When consumer desire did not keep pace, the need for a waste solution remained acute.

The history of midcentury efforts to redefine dairy waste offers new insights into how consumerism and health mediated human relationships to nature. Skim milk's transformation from hog slop to desired diet food came about not because of demands from consumers for low-fat food, but to solve a manufacturing waste problem. Yet consumers were not irrelevant to the new classification of dairy waste as a valued product. Consumers' pursuit of healthful landscapes eventually led dairy manufacturers to promote skim milk as a nutritious food. Some environmental historians studying the role of health in environmental politics have paid close attention to the role of sanitarians and public health reformers in remedying polluted waters in the Progressive Era. Others have noted the importance of health to postwar recreationists.[141] Efforts to clean up whey and skim milk demonstrate a case in which sanitary experts' and recreationists' pursuit of healthful environments intersected; by the 1950s, conservationists worried about fishing and canoeing joined sanitarians to strengthen enforcement of water pollution laws.[142] In the same way that consumers' quest for pure milk altered production processes in the Progressive Era, distaste for pollution drove changes in dairy production at midcentury.

In an industry ever more dependent on marketing even the detritus of dairy waste, appealing to the natural origin and substance of dairy foods fell flat. Whey powder, dried milk, and casein were not untouched, essential, authentic, "natural products" but unnatural, highly mediated, industrialized ones. Calcium lactate or whey protein was not a food harvested "direct from nature," but one processed, packaged, and incorporated into other foods. Such products symbolized technological transformation; they stood for the process of perfecting nature, not nature itself. Herein lay the challenge for dairy manufacturers by the mid-twentieth century.

On one hand, dairy companies were committed to making milk production more controlled and predictable. Scientific innovations that reduced natural variability enabled greater productivity. A farm utilizing AI was less susceptible to disease and bull attack and more likely to boast high-yielding cows. A factory equipped with skim milk drying equipment could send more products to market.

But, on the other hand, commitment to these modern techniques introduced new challenges to characterizing milk as a pure, unadulterated product of nature. This disconnect—between the changed and modernizing nature sought by those on the dairy farm and the imagined vision of the farm held dear by consumers—would only deepen after World War II.

From the Ice Cream Aisle to the Bulk Tank

The Postwar Landscape of Mass Consumption

Around 1950, two American families gathered for kitchen-table chats to chart the future. On the Lower East Side of Manhattan, Cele Roberts, pregnant with her first child, discussed with her husband the ideal location for raising a family. They had visited friends in some of the new houses appearing on the city's edges and swooned at the gleaming kitchen appliances the homes held—double stainless-steel sinks, new GE stoves and ovens, and Bendix washing machines. In 1949, the Roberts family purchased a house in Levittown, New York, funded by a Veterans' Administration mortgage and Mr. Roberts's income as a substitute teacher. Compared to their cramped and cockroach-laden one-room apartment, the eight-hundred-square-foot house felt spacious. The Robertses' external surroundings also differed from the "hurly-burly" of the city streets of their previous home. Mrs. Roberts could peer out her picture window and track the growth of newly planted fruit trees and beyond them a "long grassy hillock."[1] At about the same time, in another household, an Illinois farmer named Dale Beall shared his sketches of a remodeled dairy barn with his wife Veleanor and their daughters. Like the modern kitchen of Levittown floorplans, the new milking parlor and bulk tank cooler in the Beall barn plan carried hopes of a brighter future. By streamlining the architecture in which their cows were fed, housed, and milked, the Beall family aimed to bring the old family farmstead into a new era. "Fifty-six," Mr. Beall told his daughters, "is the year to fix."[2]

Although the Roberts and Beall families rebuilt their lives during the same period, their stories rarely intersect in historical works. Those who moved to suburban neighborhoods often perceived the areas to which they moved as empty spaces, blank slates on which they could rewrite their lives.[3] Others adhered to an idealized pastoral view of the rural landscape as an unchanging place. Both these ideas masked the ways in which the nation's rural landscapes were changing and, with them, the lives of those whose incomes derived from farming. Historians and geographers have offered an admirably detailed picture of how supermarkets, strip malls, and suburban tracts of ranch homes remade the built environment of the postwar United States. But as the Bealls' farm plans suggest, the building blitz in the suburbs was paralleled by a grand reconstruction of the nation's barnyards. The logic of mass consumption that altered the sites where Americans lived and purchased food and other goods simultaneously transformed the geography of food distribution and the agricultural infrastructure.[4]

Tank trucks, traveling on the same interstate highways that facilitated suburban development, made it possible to haul milk once deemed too expensive to get to thirsty urban and suburban consumers at lower costs.[5] Thus, geographically isolated farmers who had once sold milk for manufacturing purposes (to cheese factories or creameries) could sell milk on the Grade A market. But to sell their milk for drinking, outer-ring dairy farmers had to remodel their dairy barns and milk houses to meet the more stringent health standards required for fluid milk. When the Beall family added a milking parlor and bulk cooler, their remodeling job was part of this broader effort. Once farm families remade their barnyards, their milk could be channeled into a greater variety of foods and reach a broader number of consumers; it might end up at the nearby plant that made award-winning cheese, but it could also be served pasteurized in a glass on the kitchen table of a suburban development like Levittown. By 1959, the milk of farm families who supplied Wisconsin Consolidated Badger Cooperative was reaching consumers at all of the Carvel ice cream stores in Milwaukee, as soft-serve in Dairy Queen and Tastee Freez stores in Michigan and Wisconsin, and in custard stands throughout the state.

Milk processors welcomed the greater volume of milk that could be diverted to whichever product promised the greatest reward, but realignment of the milk commodity chain had mixed effects. As their sales expanded geographically, milk cooperatives like Wisconsin's Badger boasted fewer members. Many farm families who believed they might reap greater profits by remodeling their barns to meet Grade A standards found that the cost of improvements outweighed any premiums paid for improved milk. Farmers who had long produced Grade A milk faced competition from distant counterparts that drove milk prices lower, even as housing developments on the urban periphery nudged property tax bills upward. Consumers could obtain a broader array of dairy products at the supermarket, but traded some control over their diets to food manufacturers. The rise of a food economy of mass consumption and production was not smooth and foreordained, but was fraught with discord and dislocation.[6]

Both newly suburban residents and the farm families that stocked their refrigerators made personal decisions about how fully to engage with the postwar food economy. Not all farm families remodeled their barns to stay in the dairy business. Not all urban residents aspired to a suburban existence where they could fill their refrigerators and freezers with foods purchased from the nearest supermarket. As changes in how foods were marketed, packaged, and transported redrew the geography of food production and consumption, however, farm families and consumers could hardly avoid the impact of these changes on how they sold their product or bought their food. Through the story of ice cream and its ingredients, this chapter examines how individual farm families, milk inspectors, and consumers navigated the broadly changing postwar landscape of dairy production and mass consumption.

We All Scream…

Ice cream showed up in nearly every iconic postwar locale. Baskin & Robbins and Carvel were both founded in the immediate postwar era, and by 1954, 40 percent of all incorporated towns over 2,500 had a Dairy Queen.[7] Vendors like Good Humor sold their wares on leafy suburban avenues, beckoning residents and especially children outdoors for a frozen treat. According to architectural historian Dolores Hayden, developers seeking to attract buyers to the suburbs even used a house's location on "the regular route of the Good Humor Man" as a selling point.[8] Ice cream appealed across racial and class lines; features and ads for ice cream desserts appeared in *Ebony* magazine, as well as in mainstream journals.[9] With the widespread adoption of home freezers, more Americans in the 1950s and early 1960s could take home a half-gallon or even a full gallon of ice cream from the supermarket.

The site and character of ice cream consumption in the postwar era differed from its prewar patterns. From the 1910s through the early 1940s, Americans usually purchased ice cream from confectionaries, drug stores, and retail counters, since most Americans lacked home freezers to keep ice cream cold. Eaten outside the home, ice cream was intertwined in patterns of community life and social relations. Rural residents caught up with neighbors in town over a malt or milkshake on Saturday nights. Children chased the local ice cream man down public streets. In the lyrics of tunes like "Over a Chocolate Sundae on a Saturday Night," and "Won't You Have an Ice Cream Soda with Me?" romances blossomed at the soda fountain counter.[10]

When Americans of the interwar period *did* serve ice cream at home, most of them purchased only as much ice cream as they planned to eat, since they had no way to store it for future use. In 1940, only about half of all American homes had electric refrigerators.[11] Thus, ice cream was a food of picnics, banquets, birthdays, and special events. Women could purchase ice cream pies for special guests, turkey-shaped desserts at Thanksgiving, or animal-molded sundaes for a birthday party.[12] At summertime picnics, families made their own ice cream. One

New York farm family celebrated the Fourth of July in 1943 with an ice cream social.[13] Similarly, making ice cream together was *the* event at a 1941 Maine family reunion. Men dug out a block of ice and pounded it into small pieces with an axe, while women mixed a custard of cream, eggs, and sugar. Finally, everyone took turns cranking the makeshift "freezer" in the yard.[14] The labor involved in making ice cream marked it as a food of summertime gatherings that were much anticipated precisely because they were so infrequent and unusual.

After World War II, however, ice cream consumption changed dramatically. Ice cream consumption hit an all-time high in 1946 at 5.1 gallons per capita.[15] Not only did Americans eat more ice cream, they also purchased it in new places. Three factors were especially important in these changes: the adoption of home freezers, the demographic shift of the baby boom, and the rise of supermarket shopping. First, more consumers purchased a refrigerator with a freezer compartment in the 1950s. Not only were refrigerators readily available, but postwar prosperity and installment plan purchasing made the products afford-able. In the five years after World War II, consumer spending on household appliances rose 240 percent. In 1950 alone, Americans purchased three million refrigerators, many of which came with freezer compartments.[16] Once American households had home freezers, it became possible for their residents to purchase ice cream on a more regular basis and in larger quantities.[17] Whereas in 1941, half-gallon and gallon packages made up less than 1 percent of the ice cream market, by 1955, nearly 30 percent of the nation's ice cream was purchased in half-gallons and gallons.[18] The debut of the television set further stimulated the growing home market for ice cream.[19] Ice cream and appliance makers fully rec-ognized the mutually beneficial elements of their relationship; in a 1950 ad campaign, freezer and refrigerator-maker Kelvinator sought to boost sales by promoting ice cream during June dairy month.[20]

The postwar baby boom also played a key role in changing the patterns of ice cream consumption. In the late 1950s, ice cream manufacturers began selling ice cream sandwiches and other novelties in multipacks, capitalizing on the growing size of the typical American family. The crowds of kids flocking to the nearby ice cream stand or chasing the ice cream man down the street were a ripe market. As the decade progressed, ice cream makers appealed to children in increasingly targeted ways. Eskimo Pie awarded children who sent in ice cream wrappers with premium gifts, like roller skates, a Joe DiMaggio baseball, and pen and pencil sets.[21] Carvel Ice Cream Company Stores interested kids by creating "Freezy the Clown," staging ice cream eating contests at new branches, and developing a comic books series starring Freezy Carvel, Super Space Man.[22] Due to its ease of preparation, ice cream was well suited to an age in which women still shouldered the primary responsibility of making meals and caring for chil-dren but joined the paid labor force in greater numbers than ever before.[23]

Figure 4.1 Ice cream takes on a patriotic appeal in this image by the Olsen Publishing Company. Used to boost troop morale in World War II, ice cream consumption reached record highs in the late 1940s. Wisconsin Historical Society, WHi 93317.

As home freezers facilitated the ice cream binge and the baby boom increased consumer demand, supermarkets emerged as a key site for ice cream purchase. In 1952, though the nation's 15,383 supermarkets made up less than 4 percent of grocery stores nationwide, they conducted over 40 percent of the grocery business.[24] The food store came within 1 percentage point of the drug store as the place where Americans bought ice cream in 1950.[25] Buying ice cream in a supermarket was fundamentally different than purchasing it from a confectioner or drug store. The heart of the supermarket sale was impulse buying; sales managers arranged products on the shelf in attractive displays in hopes of sparking unplanned purchases.[26] Consumers adjusted slowly to the idea that ice cream could be purchased on impulse and kept in the freezer for unplanned events. A 1954 marketing survey of 193 housewives in Philadelphia and Pittsburgh determined that women over forty, in particular, were more likely to purchase ice cream for one meal only. But the same study revealed that younger women, especially those with children, purchased "extra quantity for storage in refrigerators."[27] Products like Eskimo Pie's multipacks were specifically designed for supermarket distribution. To ensure that its growing line of ice cream bars appeared in large stores, Eskimo Pie even designed its own merchandising cabinet.[28]

Access to such an abundance of sweetened, frozen foods marked a radical break from the rationing of cream and sugar during the war years. But the dizzying array of frozen desserts in the supermarket aisles and flowing from fast-food shake machines could also be disorienting. In 1954, the executive director for the Detroit Dairy Council, having interviewed one hundred consumers, told the Michigan Association of Ice Cream Manufacturers that consumers felt less sure about what, exactly, filled the cones and cartons in the freezer case. She explained, "In interviews and contacts with consumers, there permeated the feeling that ice cream manufacturers were selling something called ice cream that contained substances other than milk, cream, sugar, and flavoring."[29] The same technologies and economy that made it possible for ice cream to become a food of all seasons and incorporated in the diet of all classes led some to wonder whether such a product might be cheapened in the process of mass access. How good could ice cream be if manufactured on a mass scale?

The sources of consumers' questions about ice cream reflected broader concerns about food in the 1950s. Many consumers' biggest concern was value. Inflationary pressures in the 1950s hiked food prices, and food buyers wanted to get their money's worth for the goods that filled their grocery cart. Historically, demands for low prices centered on staple foods like milk and meat, but as Americans came to think of a wider variety of foods as needed items, they expected value to extend to goods once thought of as luxuries—including ice cream. Value entailed quality, as well as cost. Consumers were especially loath to pay high prices for substandard products and many felt that quality suffered as the assortment of frozen treats abounded. Stated a Washington, D.C., consumer, "The stuff [ice cream] sold today bears little resemblance, either in looks, taste, quality, and in all probability, ingredients, to that manufactured several years ago."[30] Paying inflated prices for poor-quality ice cream left a bitter taste in consumers' mouths.

If an acute sense of value put consumers on edge, the changing process of ice cream manufacture intensified their concerns. In the 1950s, food scientists concocted new recipes for ice cream mixes. To the core ingredients of cream, sweetener, and flavoring, ice cream makers added stabilizers, like gelatin, Irish moss (carrageenan), and guar seed gum, among others. Before the war, ice cream makers used stabilizers to help their product withstand inadequate refrigeration. Newly formulated mixes enabled ice cream to travel longer distances while maintaining its texture. Stabilizers slowed the formation of ice crystals and preserved the product's creaminess as it traveled from factory floor to supermarket freezers.[31] Manufacturers' desire to distribute a low-cost product also led some to rethink the basic components of ice cream: milk, cream, and sugar. Food manufacturers in Texas began to replace cream with vegetable fats, and these "vegetable fat frozen desserts" reached as far north as Chicago by 1952.[32] As ice cream makers utilized a greater variety of ingredients to provide a cheap abundance year-round, consumers' suspicions about what, exactly, was in their ice cream cartons only intensified.

The variability in ice cream recipes led states to revise regulations defining ice cream. State laws varied. In Arkansas, ice cream could only be sweetened with sugar, but in neighboring Missouri ice cream makers could use dextrose, lactose, corn sugar, maple syrup, honey, brown sugar, malt syrup, and molasses (except blackstrap) to sweeten their product. Alabama allowed nondairy fats—such as vegetable and animal fats—to be mixed with sweeteners and sold as "mellorine" or labeled "imitation" ice cream, while Wisconsin prohibited any iced desserts using vegetable fats that resembled ice cream to be made or sold in its state. The patchwork of state laws reflected the predominance of the local agricultural geography. Arkansas sugar growers and Wisconsin dairy farmers benefited from standards that restricted ingredients manufactured by other industries.[33]

But as food manufacturers, rather than farmers, flexed their muscle to shape food policy, they sought the creation of a new set of federal standards for ice cream to govern its contents. The supermarkets that sold the bulk of the nation's ice cream by the late 1950s and the food companies that supplied them were part of a nationwide distribution system. When food manufacturers sent their goods across state lines, their products crossed into new regions of legal authority. Unifying food standards under the federal umbrella would enable even greater flexibility for food manufacturers in obtaining raw materials and selling their final product. Food processors argued that uniform regulations would allow the logic of supply and demand, not state regulation, to drive their sales.[34]

To this end, in 1958, the federal Food and Drug Administration (FDA) issued new standards for ice cream that required any ice cream to contain at least 10 percent milk fat and 20 percent milk solids (milk fat and nonfat milk). The standard also allowed ice cream makers to utilize some ingredients as stabilizers, as long as these substances had been proven safe and did not exceed more than half of 1 percent of the finished ice cream.[35] In announcing the federal ice cream standards, FDA regulators noted that the new rules would ensure that consumers were getting the product they paid for—without excessive amounts of air, water, or deleterious substances mixed in.

Although the FDA's alleged aim was to make sure consumers obtained "what they expect in these dairy foods," the establishment of federal standards on ice cream made it more difficult for consumers to assess the foods' contents. Once the FDA established a federal standard of identity for a food product, its manufacturers were no longer required to list which ingredients it contained, so long as it met the guidelines set forth by the federal standard. A carton of ice cream might be churned entirely from cream or blended from a combination of dairy foods (like skim milk, casein, dried buttermilk, or whey), but as long as the end result contained 10 percent milk fat and 20 percent milk solids, it constituted ice cream under the federal standard. The standard thus set a basic floor for what defined ice cream and asked consumers to trust food manufacturers and regulators to look after its contents.[36]

Many consumers, however, preferred to maintain the authority to evaluate the contents of their foods. They clamored for ice cream cartons to carry comprehensive lists of ingredients. As one man explained, "When buying a can of salmon, the label tells me that SALT has been added. When buying canned fruits, the labels proclaim the addition of SYRUP. Even packaged candy-coated popcorn holds a slip stating the ingredients…but not ice cream!"[37] Some consumers wanted ingredient lists because they were anxious to know what they were feeding their children.[38] Others wanted to avoid gelatin or hog-derived ingredients for religious reasons.[39] Some wanted to select ice cream with the highest butterfat content, while others sought low butterfat desserts.[40] Consumers felt that they had few clues to distinguish between the ever-widening array of items in the freezer case. "When I pay from $0.59 to $1.59 per half gallon," stated Theodora Burke, of Utica, New York, "I would like to know what the difference is."[41] Barring ingredients lists, consumers were more likely to believe scare stories printed about ice cream's contents, which claimed that modern ice cream contained the same ingredients as antifreeze, rubber cement, and oil paints.[42]

The intense desire to know what was in ice cream was about more than changing manufacturing processes and the legal guidelines of its creation. It was also about the insecurities that home cooks felt as they furnished family tables with premade items, rather than cooked meals from scratch. To women balancing a host of responsibilities, scooping out a dish of ice cream or handing out ice cream bars was appealing because it required less time than baking pies, cakes, or cookies. Yet ice cream carried none of the cachet of a home-baked tart. In a time that glorified domesticity, many women felt that the foods they served reflected their worth. The dessert course, in particular, was often viewed as the best test of a woman's skill in the kitchen. A woman who could bake a tasty pie or an elegant cake earned admiration. But the respect conferred by a well-baked pie crust was difficult to attain when serving a manufactured product. Whereas a home baker could certify the quality of a cake's contents and knew the process of its creation, the housewife who served store-bought ice cream could offer few such assurances. One could buy ice cream and remake it at home—enhancing it with delectable sauces, decorating ice cream cones as clown hats, or layering ice cream into frozen pie crusts—but such tasks took just as much time and skill as baking. Being fastidious about the act of food purchase—that is, certifying that a product's contents were wholesome and becoming an avid reader of ingredient labels—was a way to adapt homemaking skills to a consumer age.[43]

Consumers' demands for more information about the contents of ice cream also reveal how central product labels and ingredient lists had become to food buying by the 1950s. Initially the site of brand promotion, careful attention to food labels became further entrenched as food manufacturers utilized them to bill foods' nutritional contents in the 1920s and 1930s. As food manufacturers' lexicon became more difficult to decipher, consumers demanded a right to information about food contents, leading to the creation of government-enforced quality standards for food labeling in the 1930s.[44] By requiring government quality standards, consumer advocates in the 1930s aimed to empower food consumers to make informed choices about their foods. But as practiced in the 1950s, the FDA's standard of identity ruling provided consumers with less evidence, not more, about the ingredients within a carton of ice cream. While American women willingly traded away the labor of making dessert to ice cream manufacturers, they wanted to maintain some control over the quality of foods they served in their kitchens.

The questions that ice cream inspired consumers to ask demonstrate the ambiguities of mass consumption in the postwar era. On one hand, the techniques of mass production, such as long-distance transportation and innovations in food science, made it possible for a broader number of consumers to experience ice cream as an everyday treat. The broad array of ice cream choices— from soft-serve cones to Klondike Bars, hand-packed pints to gallons of factory-churned vanilla—signaled the height of a consumer-driven economy. But, on the other hand, behind this delectable array of choice lingered a sense of uncertainty. What shortcuts made such cheap abundance possible? Could a woman be a "good" cook in an age of manufactured foods? Consumers' desire to know what was in ice cream usually led them to contemplate the process of its manufacture and especially to wonder about the additives or nondairy fats it contained. Rarely did they consider the processes by which the "natural" ingredients—like sugar or cream—were produced. Had they done so, consumers would have found that trends toward uniform standards extended beyond the ice cream recipe to the farms on which cows produced milk. There, farm families, like housewives, struggled to balance the promises of mass production with maintaining individual control over their livelihoods.

A New Recipe for the Farm: Bulk Tanks, Loose Housing, and Stored Grass

Postwar transportation technologies and political changes radically altered the geography of dairy farming. Before the war, the seasonality of milk production, transportation costs, and economic and health regulations created two zones of milk production. Farmers located in greatest proximity to the nation's urban areas provided city dwellers with Grade A milk for drinking. These Grade A producers earned a premium for their milk, because it had to be produced under more stringent health regulations and shipped to market quickly. Alternatively, those farmers located at a distance from urban markets largely sent their milk to local manufacturing plants, such as creameries or cheese factories. Grade B producers did not have to invest in the equipment to meet Grade A standards and were paid less for their product as a result. While surplus milk from Grade A farms could be made into cheese or butter, milk that derived from a manufactured milk farm could not meet the sanitary standards to be bottled and sold as fluid milk. Outer-ring Grade B dairy producers became newly able to supply the fluid milk market in the postwar era, so long as they improved their farms to meet Grade A production requirements. By flattening transportation distinctions between manufactured and fluid milk, the new postwar food economy made it possible for milk to be interchangeable.

The changing geography of milk transportation affected every link in the commodity chain of milk production—those who shipped milk, inspected it, and manufactured it into dairy foods. Small milk-processing plants that had once churned butter or crafted cheese grew into national distributors of a wide variety of dairy products to supermarkets and fast-food restaurants. According to a 1960 study of dairy manufacturing cooperatives in Minnesota, Iowa, and Wisconsin, half of cooperatives surveyed added facilities to their plants to accommodate Grade A milk between 1951 and 1956.[45] One such cooperative, Wisconsin's Consolidated Badger Cooperative, purchased smaller cheese and ice cream companies and closed some smaller butter factories, directing milk from an ever-expanding region to larger manufacturing plants. Pennsylvania's Wawa Dairy Company purchased nearby Brookmeade Dairies in 1950, and by 1957, Wawa's milk trucks began carrying ice cream.[46] By acquiring the facilities to process fluid milk, cheese, cream, ice cream mix, and dairy byproducts like powdered skim milk, dairy manufacturers could sell the dairy products in highest demand for maximum returns. The goal of these plants was not simply to provide a market for local farm families to sell their milk, but rather to obtain milk from a greater region and sell it nationally.

Milk inspection also witnessed great changes. The transport of milk across greater distances required municipal milk inspectors to rely more heavily upon counterparts from other cities or states to guarantee milk's safety, for they could not travel to increasingly distant climes to carry out inspections. In 1948, for the first time, the New York City Board of Health approved cream for ice cream manufacture to enter the city from outside its own milkshed, that is, from outside the geographic area that supplied the city's fluid milk. Some inspectors worried that interstate agreements allowed inexperienced inspectors to take over the tasks of long-serving municipal officials and insisted on asserting local control over the milk and food of their localities. But states that exported vast quantities of milk, like Wisconsin and Vermont, were especially eager to set uniform standards of milk inspection, because interstate and national agreements would enable their states' milk to reach consumers without being stymied by local regulators.[47]

Those on the dairy farm felt the effects of the changing dairy economy most profoundly. As before World War II, the kinds of farm families engaged in dairying varied widely in the immediate postwar era. Over three million farms reported milk cows in the 1950 agricultural census, but only a fraction sold milk commercially. (Nearly half of the farms reporting dairy animals had just one or two cows.) On farms that sold milk, herd sizes remained small, for dairying was often combined with general farming. In Wisconsin, for instance, most farmers had herds of ten to nineteen milk cows. Only California and New York boasted a significant number of herds of fifty or more milking animals. Even in California, the state with the largest average herd sizes, fewer than 10 percent of the state's dairy farms had more than fifty cows.[48] Since dairying continued to be mixed with other farm pursuits, most farms had buildings fitted to accommodate generalized farming, not specialized dairying. Depression economics and wartime equipment shortages discouraged new construction in the 1930s and 1940s. A 1947 survey of Illinois dairies, for instance, found that most farm buildings had been constructed forty to seventy years earlier. Such buildings were physically sound but ill-equipped to house larger herds of dairy cows or equipment specifically designed for dairying.[49]

Rather than spend the money on farm improvements to enter the Grade A market after World War II, many farm families left the dairying business. Farm income slumped with the rising costs of machinery. Some dairy farmers also faced higher property taxes as suburban neighborhoods encroached on their fields and as tax assessors recalculated the value of their farms to incorporate new buildings.[50] Nearly half of the commercial dairy farms in the central plain region of New York in 1953 were no longer in the business by 1963.[51] Minnesota lost 29 percent of its dairy farms between 1962 and 1968.[52] Many of those who left dairying stayed on their farms but shifted to raising beef cattle or cash crops, such as corn, soybeans, or sugar beets—agricultural operations that required less intensive labor or less expensive machinery.

Families who sought to continue dairying rebuilt outmoded barns and reconstructed the environment of dairy farming to achieve mass milk production in the postwar era. They rethought nearly every element of farm management and the built environment. New state and national standards for milk quality, such as the 1959 United States Public Health Service Milk Ordinance and Code, encouraged some remodeling, because the laws defined milk's purity partly by the architecture in which it was produced.[53] Some farmers sized up to accommodate larger herds and, they hoped, a bigger milk check. Others recalibrated the mix of land devoted to croplands and permanent pasture. The impact of the economy of mass consumption, then, was not only evident in the broader array of products in the supermarket aisle, but also in the ways that farmers organized their barnyards, housed their cows, and grew crops to feed them.

Just as home freezers remade the suburban kitchen, the bulk tank cooler transformed the milk house. Until the early 1950s, most farmers chilled milk by pouring it into ten-gallon cans and immersing the cans in cold water in a cement tank. When milk haulers came to pick up the milk, the farmer and the truck driver would hoist the full cans onto the hauler's truck, in exchange for empty cans that would be filled the next day. During the 1950s, dairy farmers began utilizing bulk milk tanks, mechanical refrigerators capable of chilling hundreds of gallons of milk. Rather than heave heavy cans of milk, haulers servicing bulk tanks hooked a special hose to farmers' bulk tanks and pumped milk into trucks' stainless-steel receiving vat, making the chore of milk cooling, transportation, and handling more efficient.[54] Most dairy farmers needed to build new milk houses separate from the barn to accommodate tank truck deliveries.

In 1953, only 6,200 dairy farms, or roughly 1 percent of the commercial dairy farms in the United States, had bulk milk cooling tanks. Farm people in regions where large herd sizes prevailed adopted bulk milk tanks quickly. The milk of an average-sized California dairy herd of forty-four cows, each of which yielded 8,000 pounds of milk per year would have filled eighty or ninety cans each day— too many to heave and hoist onto a milk truck.[55] Efficient for farmers with large herds, bulk tanks were costly for the smallest dairy farmers. At $1,500 to $2,500 dollars, the purchase of a bulk tank was a considerable expense. As farmer Herbert Leon of Aurora, Indiana, asked Senator Clyde Wheeler in 1956, "The talk now is bulk coolers which cost $2415. I only sold $2886 worth of milk last year

Figure 4.2 The bulk tank cooler required significant investment from farm families and indicated a farm's specialization in dairying. It was a source of pride for those who invested in the technology in the 1950s and 1960s. Consolidated Badger Cooperative Creamery Records, Wisconsin Historical Society, Green Bay Area Research Center.

how could I buy one and eat?"[56] The costs of installation, such as remaking the milk house door to accommodate a tank or installing pressurized water to sufficiently clean it, added to farmers' expenses.[57] Since the bulk tank was a new technology, there were few used models for economizing farmers to purchase.[58]

The bulk milk tank was just one element of the farm geared to mass milk production. Farm management specialists in the 1950s also promoted plans to redesign two parts of the dairy farm that had remained unchanged since the turn of the twentieth century: the stanchion barn and the pasture. Their aim was to help farmers boost production and raise more cows on the same acreage without additional labor costs. National trends to mass production thus transformed not only the networks of milk transportation but also the layout of dairy farmsteads themselves.

One change promoted in the 1950s was to shift from stanchion barns to loose housing. Until the late 1940s, most dairy cows were housed in barns in which each cow stood confined to an individual stall or stanchion. Farmers milked, fed, and bedded cows in their stanchions. The stall barn made it easy to locate specific cows. If the veterinarian came to administer medicine or if a buyer came to purchase the animal, a farmer knew where to find it. Stanchion barns, though, were expensive to construct and costly to remodel. As the average size of cows increased, farmers found the stalls too cramped or short to accommodate cows' larger bodies. Engineers believed that they could improve cow comfort, reduce farm costs for improvements, and relieve laborious barn tasks, such as feeding and manure removal by altering the layout of the stall barn.[59]

In the 1950s, engineers experimented with a different kind of dairy barn: the pen barn or loose-housing system. In these barns, used widely in the South before World War II, spaces were arranged by function: one section of the barn was used for feeding, another for loafing and bedding, and a third as a milking parlor. Adult cows were allowed to roam freely, rather than being fixed in place.[60] Whereas in a stall barn farmers provided feed to cows in the feed alley adjacent to their stalls, in a pen barn cows ate from self-feeding bunks that held hay and silage. In a stall barn, each cow had a separate drinking cup, but in a pen barn, cows drank from a common water tank in the feeding area. Farm engineers believed that loose housing would save labor and reduce barn construction costs. Pen barns were easier to remodel than stanchion barns, so they could accommodate fluctuations in herd size. The reduced costs in providing loose housing made it the favorite of farm engineers. The conclusions of one agricultural economics student in 1961 summed up their view: "If new dairy buildings are necessary, stanchion barns should not be built. There is always a loose-housing alternative that gives lower cost."[61]

Changes to the task of milking in a loose-housing system were especially dramatic.[62] In a stanchion barn, a farmer moved from cow to cow during milking, usually with a milking machine in tow. Farmers with loose-housing operations milked cows in a central location, called a milking room or milking parlor. Twice a day, cows filed into the cement-floored room in small groups and proceeded into individual milking stalls. The stalls, located above a pit where the milker stood, allowed farmers to work at eye-level with cows' udders. Thus, farmers could inspect cows for signs of disease or injury and milk cows without stooping or squatting. As the milker washed cows' udders and hooked them to the milking machine, cows ate concentrated feed or grains. In most modern milking parlors, pipelines carried the milk directly to the bulk tank. When all the cows were through milking, they exited the parlor, and the next batch of cows entered.[63]

Even farm families reluctant to adopt loose housing entirely built milking parlors and installed pipeline milkers.[64] Dale Beall was among them; he maintained a stall barn but proudly displayed a new milking parlor by late 1956. Milking parlors appealed to farm families because they cut the time required to milk cattle. For farm owners who hired laborers to perform milking tasks, the time saved with the use of a milking parlor was paramount. In 1963, Max Foster, of Modesto, California, credited the construction of a new milking parlor with increasing the

average number of cows each employee milked from sixty to one hundred animals. For Foster's operation, the main distinction was not just that parlor
machines channeled milk to the cooler more efficiently, but also that the new
machines had automated feeders that provided each cow with a grain ration in
proportion to the amount of milk produced. By streamlining milking time, the
milking parlor thereby diminished the wages that Foster owed to workers.[65]
Hank Bennett, who farmed in Sixes, Oregon, also noted the changes that the
construction of a milking parlor brought to the task of milking. In his newly constructed milk room, one man could milk and feed 150 cows, thanks to machines
that massaged cows' udders to stimulate milk to be let down and push-button-
operated conveyor belts that delivered roughage at milking time. Cows, not just
laborers, faced adjustment to the new system. After operating his remodeled
milk room for a few weeks, Bennett reported, "The heifers are now coming in
with their milk let down, but many of the old cows are only partially ready. It will
probably take several months to get some of the old girls trained."[66]

A final change to the local geography of the dairy farm in the 1950s and 1960s
dealt with how and what cows were fed. A cow's ration continued to be made up
of a mix of roughage (hay, silage) and concentrates (grains or root crops), but
the ways they received their roughage began to change. For the first half of the
twentieth century, one of the most consistent components of dairy cows' roughage ration was pasture grass. Just as consumers greeted the first warm day of
spring with an ice cream cone, farm families marked spring by turning cows out
to pasture. Elihu Gifford even recorded the precise day when he turned cows out
to pasture for the first time in his diary each May.[67] Grazing restored cows' health,
gave a rich yellow hue to their milk, and increased the flow of milk they provided.
When pastures were dried up or covered with snow, cows ate the roughage part
of their ration as hay and silage.

Dairy farm families did not abandon grass feeding for corn in the 1950s and
1960s, contrary to the impression created by recent agricultural writers and
grass-fed meat enthusiasts.[68] In fact, many dairy farmers became more reliant on
the roughage part of the ration after 1945.[69] But the ways in which farmers utilized pasturelands and cows encountered grass did undergo a change. At the very
moment that urbanites were drawn to the suburbs by visions of cows grazing in
verdant and rich pastures, farmers began to feed cows grass in the barnyard
rather than turning them out to graze. Some farmers provided grass chopped,
others as silage, and some used hay pellets or wafers. These changes to the form
in which cows received pasture rations required changes to the physical layout of
farms and to farm laborers' everyday tasks.

One strategy to intensify the use of grass involved reallocating the kinds of
lands devoted to pasture. Some farmers used electric fences to confine cows on
a particular area of pasture each day, thus preventing cows from wasting grass in

a larger area. Farmers gave grass a chance to grow back before cows grazed on it again.[70] Other farm families removed cows from upland areas that previously served as permanent pasturelands. Grazing in upland woodlands exposed cattle to dangers, like poisonous plants, and rarely proffered adequate roughage. In the 1950s, state soil conservation programs provided incentives to farmers for taking cows out of woodlands, since grazing also compacted soils and harmed young trees. Wisconsin's Forest Crop Law, for instance, reduced property taxes on woodlands from which cows were removed and thus offset the cost of building fences to keep them out.[71] As cows were restricted from forest uplands, land once utilized for growing cash crops was designated for pasturage. In New Hampshire, farmers enrolled in the Green Pastures program fenced off woodlands and planted a mix of landino clover, alfalfa, or brome grass on fields once devoted to wheat and potatoes.[72] New Hampshire's Green Pastures program was largely an effort to boost dairy yields so that eastern dairy farmers could compete with milk imported from western dairy farms, without increasing acreage. The changing national geography of milk distribution, then, instigated transformations to the layout of New Hampshire farms.

A new, more capital-intensive, way of providing roughage for cattle was to cut and haul grass to cows that remained in the barnyard. Using mechanical choppers in the field, farmers then provided the chopped grass to cows from self-feeding bunks. Those who chose this alternative believed hauling and chopping pasturage bested grazing, despite the twice daily labor involved in cutting grass. Mechanical choppers cut grass more systematically than cows, who trampled some grasses as they grazed. Feeding chopped grass on a schedule, farmers claimed, caused milk production to be more consistent. Others turned to providing pasturage in the barnyard because they found it difficult to get cows to and from pasture, especially when pastures were located across busy highways.[73] Generally, it was the efficiency afforded by cutting grass that prompted farmers to shift to this method of feeding. One Ohio farmer told *Hoard's Dairyman* in 1958 that feeding cows machine-cut pasture grass instead of grazing cows on pasture enabled him to raise forty more cows on the same acreage. California farmer Max Foster said it helped him increase his herd by four hundred animals, on only slightly larger acreage.[74]

The biggest change to grass feeding in the postwar era was its preservation as silage. Farmers who utilized this practice stored the mix of alfalfa, clover, and brome grass in glass-lined silos instead of drying the first cutting of grass as hay. Whereas hay needed to cure in the fields for days before being stored, silage could be preserved quickly, a fact with implications for cows' nutrition and farmers' hay-making process. Nutritionally, leaves remained on alfalfa stored as silage, boosting its protein content. More important to farm laborers, storing grass as silage diminished the importance of making hay "while the sun shines."

Whereas wet weather could leave hay moldy or slow its curing time, farmers might still be able to make silage in advance of a storm. Cut just before seeding, alfalfa and other grasses destined for silage would be chopped and then stored in the silo to ferment. Once stored, silage could be fed even when pastures dried up or were too muddy for cows to graze. Thus, ensiling grass gave farmers assurance of a ready feed supply no matter the weather. But such assurance came at a price; glass-lined silos, field choppers, and the automated feeding systems associated with them cost thousands of dollars, money that farm families hoped to offset with greater milk yields.[75]

Hauling pasturage to cows in the barn or feeding grass silage diminished the seasonality of cows' ration and cows' role in procuring it. Just as home freezers made ice cream less of a seasonal treat, grass silage made dairy cows' diets less variable throughout the year. At first, a seasonal pattern continued to shape the timing of grass fed as silage. Those who ensiled grass began feeding it to cows in the fall, as pastures dried up.[76] By the 1960s, farmers began using stored grass year-round, replacing cows' time on the pasture with silage and hay rations in the feedlot. Still novel in the late 1950s, farm people worried how the shift away from any pasture grazing might affect a cow's health. But *Hoard's Dairyman* reassured readers that cows without access to pasture and fed stored silage and hay exclusively on the Wisconsin Experiment Station farm "produced very well and seemed normal in every respect."[77]

Yet such a feeding strategy was abnormal in some respects, both to cows and to those who tended them. Farmers who fed cows grass silage in the feedlot reduced the initiative and control that cows themselves had over what they ate. Cows encountered grasses blended together in the feed trough, including those they did not relish on the pasture. Indeed, one of the benefits often cited by advocates of hay chopping and feeding in the barnyard was that cows could not be so selective about which grasses they consumed. Like a crafty housewife who mixed vegetables or cheap meats into the casseroles for their families, farmers' carefully blended ration cut costs and reduced cows' ability to choose what they ate.

Feedlot feeding also altered the character of farm work, a fact noted even by N. N. Allen, the agricultural expert who promoted feedlot feeding to skeptical *Hoard's Dairyman* readers in 1958. He wrote, "I'll never forget the mornings as a kid when I went out before sunrise to bring in the cows, barefooted and pants legs rolled up. The eastern sky was a blaze of color.... Birds were serenading from all sides, and I almost hoped the cows would be at the farthest corner of the pasture so it would last longer. What does lot-feeding offer as a substitute for that?"[78] Allen offered his comments as an aside, and yet his memories pointed to a fundamental change in the character of dairy work. Making grass feeding more efficient to maximize the number of cows milked per acre left little room for experiencing birds and skies in a pastoral scene.

The pastoral vision for which Allen nostalgically pined was the same idealized landscape that drew many suburbanites to the countryside in the 1950s and 1960s. But what these newcomers rarely understood was that the place they encountered was a place of modernization and mass production, not a landscape locked into a static past. Some of the cows grazing on the pasture, viewed by onlookers as pastoral symbols of a more relaxed life, were experiencing their own daily grind, commuting to the milking parlor and punching the clock twice a day at milking time. The economy of mass production that helped bring about the construction of Levittown simultaneously remade dairy farms and changed the lives of those who inhabited them.[79]

The built environment of agriculture, then, offers clues to the consequences of a mass consumption economy on nature just as revealing as the strip malls and split-level homes on the city's fringes. To enable their milk to be nationally marketed, farmers rebuilt the milking rooms, dairy barns, and feed storage structures. Simultaneously, they remade the working landscape of the farm, removing cows from woodlands and turning land once devoted to cash crops into regularly harvested hay fields. The nature to which city dwellers fled was not just one of timeless respite, but one touched profoundly by technological change and economic modernization.

Too Much of a Good Thing?

By the late 1950s, although consumer spending continued to drive the national economy, some Americans began to express misgivings about a public policy premised upon consumerism. Postwar critiques of the affluent society often focused on the excesses of suburban life. Betty Friedan illuminated the silent disparity of suburban women living unfulfilling lives. Vance Packard's *Waste Makers* took aim at the flamboyant tail-fins on cars parked in cul-de-sacs.[80] Problems sparked by excess were never those of suburbanites alone. Though often framed as a problem of postwar America's identity or spirit, the emphasis on abundance had material consequences for the nation's rural landscape and the bodies of its citizens.

Even staple foods like milk, and the farmers who produced it, were touched by crises of too-much in the 1950s and 1960s. One such problem was a surplus of milk itself. Farmers adopted techniques like artificial insemination and devised new feed rations because they believed that high-producing cows would yield a bigger milk check. But as milk yields increased, dairy products accumulated and prices remained low. In the pages of farm journals, debates raged over whether the milk surplus was the fault of Grade A or Grade B producers and how the surplus might be reduced, but the underlying message was clear: production gains were not translating into greater farm income.[81]

A second crisis of abundance facing the dairy industry was that of manure. Before the postwar era, most farmers handled the manure produced on their farms by returning it to cropland, thereby improving soil fertility and crop yields. As chapter 2 details, manure was one of the products of dairy cows most valued by farmers like Elihu Gifford. But as farmers altered their feeding strategies to raise more cows on the same acreage, the volume of manure their herds produced increased. Further, as farmers shifted to providing feed to cows in the feedlot, rather than grazing cows on pasture, the manure cows released concentrated in a specific area. In the barnlot, growing piles of animal waste attracted flies and generated odors. Confinement livestock operations commonly produced more nutrients than the soil could absorb. Crop farmers found chemical fertilizers cheaper and easier to use than animal manure, reducing their desire to use the excess animal waste as fertilizer. Selling dairy manure to crop farmers required multiple steps of processing and transportation, an unattractive prospect for livestock raisers and one that boosted the price of composted manure relative to chemical alternatives.[82]

As dairy farmers retooled barns and milking parlors, then, they also incorporated new equipment to handle manure. Many replaced the hand labor of shoveling and hauling manure with timed systems that diluted manure with water, channeled it into a storage tank, and then distributed the effluent through irrigation lines on surrounding fields.[83] Optimally, the construction of manure lagoons allowed farmers to store manure until a time when it could be spread safely on fields, while protecting neighbors from the odiferous stench of animal sewage.[84] Newly installed manure systems often presented problems in practice. California farmer Ernest Greenough found that the slop washed out of the barns on his dairy farm was too thick to flow through irrigation pipes and that the watery, foul mixture tended to pool outside corrals during the summer, drawing flies.[85]

The environmental and public health risks posed by excess manure surpassed the nuisance posed by flies and odiferous vapors. In northern dairy districts, soils froze during the winter and prevented manure spread on fields from being taken up by soils. Thus, manure applied during the winter could quickly be carried away by heavy spring rains into waterways, increasing the streams' particulate matter and reducing the amount of oxygen available to aquatic plants and animals. Even when applied at the appropriate time of the year, if farmers spread nitrogen-rich animal waste too heavily on soils, not all the nitrogen it carried could be taken up by harvested crops. Excess nitrates contaminated groundwater. Already by the late 1960s, studies revealed that 42 percent of rural wells in Missouri contained nitrate in higher levels than the recommended tolerance for infants. Those wells in proximity to livestock feedlots were more likely to carry nitrates.[86] Manure lagoons could also pose problems to the farmers who operated them. If farmers lacked adequate ventilation as they removed decomposed sludge from manure lagoons, they risked exposure to toxic gasses like ammonia, methane, and carbon dioxide.[87] Problems with manure remained largely an industry concern through the 1960s. Only as environmental pressures and state regulation intensified in the late 1960s and early 1970s did many farm families utilizing confined feeding develop plans to control waste.[88]

A more public problem of abundance facing the postwar era dairy industry was new research about heart disease. Heart disease became the nation's leading cause of death in 1921, but as rates of the disease increased between 1940 and 1955, it gripped the public's interest.[89] Attention to the disease only increased in September 1955, when President Eisenhower suffered a heart attack.[90] Researchers differed on what caused heart disease, some attributing it to rising occupational stresses, others believing that sedentary lifestyles were the largest contributing factor.[91] But a third explanation, one that provoked the most concern for the dairy industry, deemed the American diet, and especially its fat content, responsible for the deadly condition.

In the mid-1950s, Dr. Ancel Keys concluded that the number of calories from dietary fat was the most meaningful factor distinguishing rates of heart disease in the United States from other nations.[92] The arteries of heart disease sufferers also pointed physicians to the dangers of dietary fats. Oily deposits lined the arteries of heart disease patients, occluding and narrowing them. Since the American diet was rich in cholesterol, and cholesterol was found in the arteries of heart disease sufferers, cholesterol-rich fatty foods seemed suspect. By 1954, specialists attending the World Congress of Cardiology asserted "high-fat diets, which are characteristic of rich nations, may be the scourge of Western civilization."[93]

The health concerns raised by heart disease in the 1950s were very different than those associated with milk in the preceding half of the twentieth century. For the first time, Americans questioned milk's healthfulness not because of its propensity to spoil or carry communicable diseases, but because its chemical makeup seemed detrimental to human health. The new focus on milk's fat content broke radically from nutritional science promoted in the 1920s and 1930s. During this interwar period, milk's fat content had been the product's key benefit, for its butterfat carried vitamin A. By the late 1950s, as scientists linked saturated fats to higher blood cholesterol, milk's richness seemed a liability.

But the connection between dietary fat and heart disease remained disputed in the 1950s and 1960s, and physicians were reluctant to endorse a low-fat or low-cholesterol diet as a means to halt the threat of heart disease.[94] In 1960, when the American Heart Association first issued a statement that reducing the diet's fat content might help to prevent heart disease by reducing blood cholesterol and controlling weight, it did so cautiously. The association advised a shift from saturated to unsaturated fats and a reduction of fat calories, but included a caveat that "there is as yet no final proof that heart attacks or strokes will be prevented by such measures."[95] In 1961, the American Medical Association's Council of Foods and Nutrition called dietary recommendations "premature." In October 1962, the American Medical Association committee went further, declaring that the "anti-fat, anti-cholesterol fad is not just folish [sic] and futile," but risky.[96]

Already on the defensive due to plummeting butter sales and rising farm costs, dairy industry officials feared that reports linking saturated fats to heart disease would spell doom for the industry.[97] Dairy officials exaggerated discrepancies among health specialists about the relationship between dietary fat and heart disease. American Dairy Association advertisements issued in 1962 called physicians' recommendations to reduce saturated fats "a highly experimental treatment."[98] In January 1963, the board of directors of the National Dairy Council announced that its members would be undertaking weight-control diets rich in milk and dairy products. In so doing, the council members believed they were waging "an all-out assault on 'freak' and fad reducing diets."[99]

As dairy industry officials attempted to discredit the link between heart disease and dietary fats, ice cream makers came to view the weight-conscious sector of the population as a growing market. Analysts urged the industry to "get ice cream on reducing diets," stressing that because ice cream had fewer calories than many pies and cakes, it could be considered "slenderizing."[100] By 1956, ice cream companies began manufacturing lower-fat dietetic desserts, ice milk, and creams with artificial sweeteners.[101] Dietetic ice creams never made up a large proportion of the ice cream marketed for the home, but almost all of the soft-serve ice cream sold was ice milk.[102]

As the dairy industry attempted to shield itself from any connection to the rising prevalence of heart disease, government agencies walked a tricky line. Long a promoter of dairy products as healthy foods, the departments of public health and agriculture hesitated to turn their back on the heralded product. But as more and more studies tied heart disease to dietary fat, the cautious stance became increasingly difficult to uphold. On January 19, 1962, CBS television aired a documentary called *The Fat American*. The program invited leading physicians and psychiatrists to comment on the causes and consequences of excess weight. Secretary of Agriculture Orville Freeman also appeared on the program. Dismissing the studies about cholesterol as "a scare," Secretary Freeman cautioned that a dairy surplus would accumulate if concerns about dietary fats prevailed. Freeman's position was roundly criticized by a reviewer in the *New York Times* who wrote, "Economic loyalty to the farm bloc may be all very well, but it hardly would seem to take precedence over what is basically a medical matter."[103] Another man who protested to the secretary's office expressed similar criticism. The caller "said he was dismayed to hear a Cabinet officer brush aside, seemingly as inconsequential, the very substantial body of medical evidence regarding cholesterol and fatty diets. He had inferred from what you said that you were more concerned about the butter industry than the nation's health."[104]

Two features made Freeman's position—calling the research linking dairy foods to cholesterol as a "scare"—increasingly difficult to sustain by the mid-1960s. First, whereas in the 1950s and early 1960s, the American Heart Association and the American Medical Association had been at odds about the prudence of recommending dietary changes to reduce the risks of coronary disease, by 1965 both groups endorsed reductions in fat intake for young men vulnerable to heart disease. As new evidence from the Framingham Heart Study revealed clearer links between blood cholesterol levels and heart disease, more physicians were convinced of the merits of dietary changes to reduce cholesterol.[105] Hence, it became increasingly difficult for the dairy industry to portray physicians' findings about heart disease as "preliminary" and their dietary recommendations as "quack diets."

The other feature that made it difficult for Freeman to uphold the importance of dairy consumption in the early 1960s was that the structural changes in dairying in the 1950s and early 1960s had altered the public perception of dairy farmers. By the 1950s, the favored place given to dairy products during World War II seemed anachronistic. Some bristled at the idea of granting price supports to farmers who seemed more like successful businessmen than working laborers. As farm people adopted new technologies and modernized their farms, retailers and advertisers of dairy foods, like the Good Humor Man, had replaced dairy farm families as cultural icons associated with dairy's life-giving properties. While policymakers in states with many dairy farmers, such as Minnesota, New York, and Wisconsin, were alarmed about the very real economic strains facing dairy farm families, such problems seemed secondary to larger concerns about the Cold War, consumer culture, and civil rights on the national agenda.

To dairy farm families and milk plant managers accustomed to a favored spot in the popular imagination, the shift away from thinking of milk as healthy and natural came as quite a shock. At the 1962 annual meeting of the Consolidated Badger Cooperative, George Rupple, the cooperative's manager, commented on the dramatic change, saying, "Today we are hearing lots of new words—cholesterol, unsaturated fats, etc. These things are affecting our lives and our business. I never thought 'nature's most perfect food' would be used as the target of health associations, faddists, and profit-minded publicists. Some people have forgotten that milk is nature's own method of sustaining life."[106]

Strangely, though, even as butterfat shifted from being one of dairy's virtues to its biggest vice, ice cream retained much of its appeal. When people sat down to a bowl full of ice cream or purchased an ice cream cone, they did so to escape and evade everyday pressures, regardless of the health infractions such indulgences may cause. Even the president's own heart disease specialist, Dr. Paul

Dudley White, was known to drink milkshakes for lunch and eat ice cream.[107] Photographers snapped his most famous heart disease patient eating a Good Humor Bar.[108] Ice cream's continued success also stemmed from another feature unique to the food. Unlike milk or butter, few Americans had ever eaten ice cream solely for its healthful qualities. Rather, Americans ate ice cream as a treat and to celebrate social events. Rarely conceived of as a health food (despite industry promotions as such), ice cream's reputation was less tarnished than that of other dairy foods by news about heart disease and cholesterol.

The resilience of the ice cream trade in the face of new concerns about heart disease, however, did not mean that the dairy industry would be unscathed by criticisms in the postwar era. In fact, by the early 1960s, research linking dietary fats to heart disease was but one of the challenges facing the dairy industry. Even more troubling was new evidence that substances created by modern society—especially antibiotics, radioactive fallout, and pesticides—were tainting the milk children drank. Unlike ice cream, fluid milk had long billed itself as a healthy food and was considered by most Americans to be essential for children's diet. If Americans turned to ice cream for good humor, they sought out milk for its nutritious, life-giving qualities. How would milk drinkers respond to the new discoveries of environmental hazards in milk? How would dairy farm families protect the ideas of healthfulness associated with their food?

5

Reassessing the Risks of Nature

Milk after 1950

On October 26, 1956, twenty-eight St. Louis women demanded information on radioactive contamination in their city's milk supply. "It appears that as we feed our children milk today, we are also feeding them radioactive materials....We believe that it is the responsibility of the Health Commissioner to make an immediate test of the milk being distributed in this area to ascertain its strontium-90 content."[1] The women were acutely aware that blasts of nuclear weapons at test sites lofted radioactive residues, such as strontium-90, cesium-137, and iodine-131, into the lower atmosphere. Winds carried these residues miles from the initial test sites. Particles of the radioactive debris settled with rain and snow. When cows ingested pasture grasses coated with this residue or grains that absorbed it from the soil, they passed a small portion of the residues to milk-drinking citizens.[2]

Inspired by concern for their families, the St. Louis women took their concerns to the Senate subcommittee on disarmament and pressed congressional representatives. As one of them explained, "As a mother, a consumer, the one most concerned with protecting the health of my family, I ask myself what can I do to protect the health of my family? Should I eliminate milk in the diet of my growing sons?"[3] In the heat of the Cold War, allusions to fallout-tainted milk became a favorite way for many—not simply antinuclear activists—to make the implications of nuclear geopolitics hit home. In 1956 and 1964 alike, presidential candidates Adlai Stevenson and Lyndon B. Johnson challenged their opponents on nuclear testing by warning of the hazards of contaminated milk.[4]

Indeed, not since the Progressive Era had milk's healthfulness and safety been so challenged as in the late 1950s and early 1960s. Particles of radioactive material were just one substance that sparked anxieties among postwar milk drinkers. Antibiotics, used to treat cows for veterinary diseases, persisted in milk and triggered allergic reactions in some who drank it. Crop-dusting planes killed crop-destroying insects but left behind chemical residues.

In many respects, postwar worries about a contaminated milk supply hearkened back to turn-of-the-century campaigns for milk's purity. Particles of stron-
tium-90 and iodine-131, like the bacilli of typhoid or bovine tuberculosis, were
invisible to the human eye.[5] Like Progressive Era dairy farmers who unwittingly
introduced bovine tuberculosis into their herds, parents feared they might un-
suspectingly feed children milk contaminated with the potentially harmful substances. Even the imagery of the campaigners against nuclear testing—a bottle
of milk with skull and crossbones—matched that of the organizations fighting
against the spread of bovine tuberculosis in the 1910s.

Despite these parallels, postwar concerns about milk's purity were distinctive.
Whereas turn-of-the-century milk reformers worried primarily about spoilage
or disease, their postwar counterparts focused on the threats to milk posed by
human-created contaminants. Indeed, some of the very technologies once credited for correcting milk's natural deficiencies became suspected sources of its contamination. Further, after World War II, consumers were increasingly aware that
milk was not purely natural, but deeply intertwined in the broad technological
changes inherent in postwar society. Milk had, of course, long been a hybrid of
cultural ideals, technologies, and environmental phenomena. But whereas citizens had once welcomed milk's technological transformation as a tool for its
improvement, postwar consumers scrutinized the social, environmental, and
health effects of science and technology in altering milk's material form.

The heightened attention to technological threats to purity carried important
implications for how and by whom impure milk was monitored, measured, and
overcome. In the Progressive Era, local authorities established and enforced standards for pure milk for drinking. Consumers exerted influence by publicizing the
results of local farm inspections. Farm people were thought essential to ensuring a
safe and clean milk supply by cooling the food and sanitizing equipment properly.
By the 1950s, the increasing scale of milk production and changing character of
contamination altered this responsibility. Many local milk inspectors lacked juris-
dictional authority, sufficient training, or equipment to determine whether milk
carried pesticide residues or radioactive particles; consequently, local inspectors
increasingly relied on federal agencies to test milk. Consumers, who previously
monitored local milk, advocated congressional action, such as the establishment of
zero tolerances for pesticide residues. Once seen as key partners in ensuring milk's
purity, farm people had less immediate control over defining and maintaining its
safety, for they could not protect cows from contaminants introduced by off-farm
sources, like fallout or pesticide drift.

A closer look at postwar efforts to rid milk of technological contaminants
enriches existing histories of the development of postwar environmental history.
Seeking pure milk was another avenue through which postwar Americans contended with the consequences of modern technology. As wilderness advocates

restricted motorized transit in natural areas, and antipollution activists promoted policies to check industrial and municipal wastes, pure milk advocates sought to eliminate technological residues in their foods. Yet pure milk was not simply milk stripped of human influence. Pure milk, like wilderness, *required* human and technological intervention—from the moment it was extracted from the cow to its storage in home refrigerators. The idea of milk as an all-natural food, pure by its nature, and threatened by modern technology, was fundamentally different than that of the Progressive Era and could only come about once pasteurization and refrigeration ridded milk of many of its perceived "natural" risks. This new vision took hold after World War II because of new technological developments and also because of broader changing cultural understandings of human relationships to the environment.

The new regulations and methods used to keep milk free of residues restructured how farm families and consumers came to know nature and evaluate the healthfulness of foods. Farmers calculated when to medicate a cow, spray their crops, or bring their cows in from the pasture with an eye toward minimizing residues. Milk shoppers read monthly reports published by the U.S. Public Health Service to find out how much strontium-90 their milk carried. This account considers how the newly emerging consumer regulations affected human relationships toward the farm environment, and in so doing suggests important ways that consumer culture shaped nature.

Finally, this account details how farm people addressed and articulated concerns about synthetic residues in the milk supply. Historians have noted that a concern for food safety galvanized broader interest in environmental issues. Rachel Carson bolstered her case for an end to the indiscriminate use of insecticides, for instance, by publicizing the chemicals' ubiquitous presence on the dinner table. Only rarely have farm producers appeared in these environmental history narratives, except as opponents of regulatory regimes governing farm chemicals. In fact, many dairy farm families expressed ambivalence about the misuse of farm chemicals, for their livelihood depended on keeping milk free of food adulterants. Incorporating farm people's perspectives helps challenge a narrative that pits industrial food producers against environmentally minded consumers.

Antibiotic-Laced Milk?: From Producer to Consumer-Oriented Regulation

Antibiotic residues were the first new milk contaminant to draw attention in the postwar era. The drugs entered the milk supply as dairy farm families adopted them to treat cows for veterinary diseases. While pesticide residues have received more attention from historians and environmentalists, the history of antibiotic residues illustrates two important shifts in the idea of milk's purity during the 1950s and 1960s: from production-oriented visions of milk's purity to consumer-driven ones, and from an emphasis on bacterial to technological contamination. At first, dairy manufacturers expressed concern about the drugs' effects on bacterial cultures used to manufacture cheese and sought farm-based solutions, much like the improved cream cooling practices they advocated in an earlier period. Over time, regulators began to examine the health implications of antibiotic residues, and manufacturers became increasingly concerned about consumers' perception of the safety of their product. Simultaneous to this shift, farm people's vision of milk's purity changed from one focused on bacterial hazards in milk to one that also incorporated keeping milk free of technological residues. Farm people's attentiveness to technological contamination would come to full fruition in the subsequent dealings with radioactive particles and pesticide residues.

Given present-day critiques of veterinary antibiotics, it is important to understand why farm people in the 1940s and 1950s turned to them in the first place. The answer lies in the swollen and irritated udders of cows infected with mastitis. As the disease pained cows' bodies, it bruised the balance sheet. Cows suffering from mastitis produced poorly flavored milk and less of it than healthy cows. Farm profits decreased as milk volume dwindled. Milk from mastitis-afflicted cows was prohibited from sale, because the same bacteria that caused some mastitis cases, *Streptococcus pyogenes* and *Staphylococcus aureus*, could also spread septic sore throat or cause food poisoning.[6] Farmers' concerns about human health and animal welfare, as well as profit-driven motives, made them determined to rid their cows of the disease.

Milking machines, adopted during the labor-strapped war years, facilitated the spread of mastitis bacteria. Like the grimy cream separators that imperiled butter quality in the 1920s and 1930s, unclean milking machines harbored bacterial communities that undermined animal health and, by extension, milk's purity. Improperly sanitized milking machine parts could transfer infectious bacteria from cow to cow, while imprecisely calibrated machines bruised cows' udders and increased the risk of infection. Vermont farmer Gertrude Mallary complained to the vendor of the milking machine, which she deemed responsible for a mastitis outbreak, "It is obvious that something about the machines, or our operation of them, is causing the trouble, because the two or three problem cows, those which were not educable to the machines, do not have these things, nor do the eight or ten which are milked by hand at the other barn.... We have too many valuable breeding animals to go on living on a keg of dynamite."[7] As modern technologies left their imprint on environmental processes, technologies like milking machines operated within broader ecological frameworks.

Pharmaceutical companies courted farmers frustrated by mastitis. The pharmaceutical industry saw great production gains during World War II and redirected their product and advertising efforts toward the veterinary market after the war.[8] They promised farmers that the drugs could restore peak milk production and improve farm revenues by ridding cows of disease. Farmers encountered advertisements touting antibiotics' healing powers and could find the drugs in many local settings. Farm journals, such as *Hoard's Dairyman* and *Dairymen's Cooperative League News*, published full-page print advertisements about the drugs. Feed dealers and farm supply stores sold the remedies. So did rural small-town drug stores, where customers were as likely to find veterinary antibiotics as Epsom salts or a soda fountain.[9]

Faced with a pressing need to heal infected cows, and the drugs' widespread availability, farmers adopted them quickly. Penicillin, the first drug certified for use on dairy animals, was the most popularly used medicine. By the 1950s, pharmaceutical companies also marketed chlortetracycline, oxytetracycline, bacitracin, neomycin, subtilin, chloramphenicol, and polymyxin to treat the disease. As the variety of antibiotics available for veterinary purposes increased, so also did the dosages contained in veterinary preparations. By 1956, experts estimated that American dairy farmers used seventy-five tons of antibiotics per year to treat mastitis.[10] Sold in one of the most common preparations—a 3.5-gram tube of ointment—this quantity would have covered the surface of more than seven football fields.

Not all were pleased with the rapid adoption of veterinary antibiotics. In the 1940s and 1950s, professionals close to farm producers—veterinarians and dairy manufacturers—voiced the strongest reservations about the drugs. Veterinarians thought the authority to administer drugs should be reserved for professionals and worried that the drugs' easy availability would lessen farm families' attention to nonpharmaceutical preventative measures—such as sanitizing the barn and separating infected cows from noninfected ones.[11]

Dairy manufacturers sounded sharper alarms about the drugs' effects on butter and cheesemaking. Cows medicated with antibiotics secreted drugs in their milk for days after treatment. These residues killed bacterial strains used to culture cheese, butter, and yogurt, thereby slowing the process of cheesemaking and marring the product's delicate flavors. By 1953, an advisor to cheesemakers in Wisconsin reported to Bureau of Dairy Industry officials that "last winter the Swiss cheese factories in the Monroe area made the poorest quality of cheese as a whole that has been made in the last twenty years....I know that the antibiotics and some of the synthetic washing powders have caused a lot of the trouble."[12] Food processors' concerns about the use of antibiotics on the farm illustrate a way that the tight relationship between farm and factory persisted in the postwar era.

Cheesemakers' problems were exacerbated by the adoption of new transportation technologies and the concentration of dairy processing, which made it more difficult for cheese- and buttermakers to pinpoint the source of antibiotic-contaminated milk. By the 1950s, an increasing amount of milk arrived at cheese factories in refrigerated bulk tanks instead of in milk cans. Milk from an entire dairy herd was pooled in the bulk cooler on the farm and then mingled with that of other farms in the stainless-steel truck tank before arriving at the factory. Thus, contaminated milk from just one or two cows on a single farm could render an entire batch of milk unfit for processing. Concentration within the dairy industry made it difficult to trace antibiotic-laden milk to its source, thereby making the relationship between farm producer and dairy plant operator more remote. Well into the 1950s, cheesemakers lacked a rapid test for the presence of antibiotics, making it nearly impossible to detect contaminated milk before combining it with untainted milk.[13]

The earliest responders to antibiotic residues, then, characterized drug-tainted milk as a production problem—one that reduced milk yields and slowed cheesemaking. Dairy manufacturers launched programs to educate farmers about the effects of antibiotics, much like earlier efforts to alter the architectural landscape of the farm to meet sanitary standards. They also pressured state and federal agencies to put tougher warning labels on antibiotics.[14] To that end, the FDA issued its first explicit regulation on antibiotics in milk in 1951, requiring that antibiotic manufacturers advise users not to market milk from an antibiotic-treated quarter of the udder for seventy-two hours. The regulation referred specifically to the impact of antibiotic drugs on cheese starters.[15]

Modern Methods from Modern Farms

Figure 5.1 The bulk tank eliminated the need to hoist heavy cans of milk but introduced new problems of milk's purity. Consolidated Badger Cooperative Creamery Records, Wisconsin Historical Society, Green Bay Area Research Center.

By the end of the 1950s, regulations of veterinary antibiotics took a new turn, from focusing on production-related measures (like the effects of drug residues on cheese manufacture) to emphasizing the potential risks of antibiotic residues on milk consumers. By that time, a growing body of evidence suggested that low levels of antibiotics in milk—especially penicillin—heightened allergic sensitivities of those who drank it. Cheesemakers, who had once framed antibiotic-laden milk largely as a production problem, incorporated the new health concerns into their efforts to convince farm families to withhold milk from treated cows. The December 9, 1959, issue of *Dairy Records* advised patrons, "If anyone is inclined to brush off the F&D regulations on the use of antibiotics treating mastitis as just another silly rule of a bogyman to harass the milk producer, he is sadly misinformed. The medical profession is in agreement that some people are so allergic to penicillin and other antibiotics that even minute quantities may bring serious consequences."[16]

The new findings on the potential health effects of antibiotic residues prompted the FDA to tighten restrictions on veterinary drugs. By 1957, the FDA established a maximum dosage for penicillin at 100,000 Oxford units, a significant reduction at a time when some antibiotic preparations contained as many as 1,500,000 units per dose. The agency also required antibiotic manufacturers to print warning labels directly on the container of the drug, not just in the literature accompanying the medicine, to make it more likely that the person administering the drugs would read the seventy-two-hour withholding rule.[17]

The FDA also stepped up testing and enforcement, especially after a simple and quick test to recognize residues in milk enabled cheesemakers and food regulators to reject contaminated milk. In 1959, the FDA launched a two-part program to sample interstate milk shipments to test for antibiotics and to train local and state milk inspectors to utilize quicker methods for residues. This program played a key role in shifting the vision of milk's purity from one focused on bacterial impurities to one focused on technological adulterants.[18]

Dairy processors and food regulators both blamed the continued presence of antibiotics in milk on dairy farmers' stubborn disregard for the seventy-two-hour withholding rule. While carelessness surely played some role, the decision of dairy farmers to disregard or uphold withholding regulations derived in part from farmers' own perceptions of milk's purity. Many of the farmers who erred in sending milk to market too soon after antibiotic treatment associated the visible healing of cows' udders with the return of milk to a suitable condition to sale. These farmers defined pure milk by the absence of bacteria, not the absence of drug residues. A dairy farmer in Kansas told the local FDA inspector that "the veterinarian had told him he could use the milk as soon as the condition for which he was treating the cow cleared up." Some farmers thought that antibiotic residues, like bacteria of milk-borne diseases, were eliminated by pasteurization.

Because the new rules on antibiotics distinguished the safety of milk from the fitness of the cow that produced it, they required a new understanding of the relationship between animal health, human health, and food purity.[19]

Further, in the barnyard, mastitis, not antibiotic contamination, posed the greatest problem. A diseased cow presented farmers with an immediate and visible challenge; antibiotic-laden milk was indistinguishable to the eye and the troubles it caused were remote. Hence, it was tempting to send milk to market as soon as a cow's udder was healed by antibiotic treatment—even if these clinical indications preceded the seventy-two hours set by FDA guidelines.

Although some farmers adjusted slowly to new ideas of milk's purity, by the late 1950s and 1960s, others were quickly becoming familiar with the perils of technological adulterants. By that time, antibiotic residues were merely one of the many potential contaminants of milk; pesticides and particles of radioactive fallout also proved problematic. Whereas fallout residues and pesticide drift were difficult to eliminate, many farmers accepted the FDA policy on antibiotic use because they could control its presence in milk by restructuring actions on the farm.[20]

Antibiotics did not eliminate mastitis from dairy herds. The disease continues to plague the nation's dairy cows. But the introduction of antibiotics to dairy herds was nevertheless important, for through their use dairy farmers became familiar with a regulatory regime that defined milk's purity by its lack of both biological contaminants and technological residues. Moreover, farmers' experiences with antibiotic residues would shape how they understood other potential food adulterants and how they evaluated regulations on pesticides and radioactive fallout.

Milk Purity in a Nuclear Age

Parents and citizens could not smell, see, or taste radioactive isotopes in their milk. Yet, between 1954 and 1965, fallout-contaminated milk became a sinister symbol of living in a nuclear age. Although worries about fallout-contaminated milk emerged at roughly the same time as problems with antibiotic residues, radioactive contamination originated from a unique ecological pathway and sparked a distinctive response. Concern about radioactive contaminants in milk originated with consumers and critics of the nation's nuclear policy, rather than from within the dairy industry itself. In the late 1950s, consumer and antinuclear organizations publicized scientific findings about the dispersion of fallout and its accumulation in human tissues. Whereas the potential health effects of penicillin were immediate, those of radioactive contaminants were sustained. The isotope strontium-90 had a half-life of twenty-eight years, meaning that it took twenty-eight years before half of its atoms decayed. As it accumulated in bones and teeth, the isotope increased rates of leukemia and bone cancer over the long term.[21] A glass of milk tainted by strontium-90, like one carrying drug residues, made palpable the inseparability of human uses of technology and environmental processes.

In the 1950s, when activists and scientists emphasized the risks of strontium-90, individual farmers and consumers could do little to track or eliminate a long-lived isotope. Farmers could not withhold milk tainted with radioactive isotopes from market, as they did with milk laden with antibiotics. But in the early 1960s, as the focus of scientists shifted to iodine-131, an isotope with a shorter half-life, farm people and individual consumers began to implement strategies to limit exposure to fallout contamination.

Studying the effort to rid milk of radioactive residues calls attention to the distinctive ways that environmental and consumer activism informed one another. The fight to get fallout out of milk suggests that consumer-oriented elements of the environmental agenda had deeper roots than often acknowledged by environmental historians, who view the consumer influence on environmentalism as a particularly post–World War II phenomenon.[22] Environmental activists drew on a longer history of consumer activism to articulate connections between efforts to combat nuclear fallout and concerns about pesticides.[23] Following the historian Lawrence Glickman, who sees the immediate postwar period as a critical but underappreciated era for consumer organizing, the 1950s were a key time for laying the groundwork for an environmental agenda that emphasized consumer protection in the 1960s.[24]

No actions better reveal the intersections between the consumer and environmental movements than those of St. Louis women involved in that city's Committee on Nuclear Information (CNI). Gertrude Faust, the chief author of the 1956 letter to the St. Louis health department that opens this chapter, was president of the St. Louis Consumer Federation. The group was a successor to the St. Louis Consumer Council, which, like two hundred others, started in the New Deal to advocate for fair prices and to counter criticism of federal price-setting mechanisms.[25] Like those who insisted on accurate food labels and fair prices in the interwar period, Faust and her colleagues sought more information to guide food-purchasing decisions. Her consumer-driven orientation was readily evident in testimony before the Senate subcommittee, in which she asked senators, "Are we to cut milk out of our diet and out of the diet of our children?" To Faust and others steeped in a tradition that sought to ensure nutritious foods at fair prices, a diet free of milk sounded preposterous. Humans could hardly abandon their need for healthy food, she argued, but they could halt nuclear testing.[26] Another who testified was Edna Gellhorn, former president of the League of Women Voters of St. Louis. Gellhorn objected to continued nuclear testing, saying, "Our children are not white mice or guinea pigs. We do not want them to suffer in an experiment to find out how much strontium-90 the human body can tolerate." Gellhorn's characterization of children as nuclear "guinea pigs" echoed the tile of Arthur Kallet's *100,000,000 Guinea Pigs*, a 1933 book that galvanized consumer activism.[27]

By the late 1950s, Gellhorn and Faust were not the only ones voicing reservations about the effects of nuclear testing. As testing accelerated, so did antinuclear activism. In 1957, a number of grassroots groups unified to form the National Committee for a Sane Nuclear Policy, or SANE. The group focused on ending the bomb tests, and by mid-1958 it had attracted 25,000 members.[28] In 1961, Women's Strike for Peace joined SANE and others in opposing nuclear testing and drawing attention to the risks of fallout with slogans like "pure milk, not poison."[29] Mainstream news outlets printed stories about fallout in food. [30]

The decision of consumer and antinuclear organizations to center on fallout-contaminated milk was strategically designed for its time but also revealed the influence of federal agencies and was rooted in the long history of the consumer movement. The risks posed by fallout-tainted milk resonated within the cultural context of the Cold War and the baby boom. Milk's characteristics—its freshness, whiteness, and important place in the diet of children—stood for purity. In the culture of the Cold War, parents worried about feeding a contaminated substance to their children and were troubled by evidence that fallout could cause genetic mutation and infertility. Spread through food, radioactive residues could penetrate what had become the very symbol of the American way of life: the suburban home. To a population at the height of the baby boom, the idea that fallout might cause genetic mutation or infertility worried parents and those wishing to have children.[31]

Emphasis on milk was also a consequence of federal procedures for monitoring radioactive substances. In 1957, the U.S. Public Health Service and Atomic Energy Commission began testing foods for radioactive content. The agency chose to focus on milk because the food was available nationwide, year-round, and could be easily sampled. In five cities—Sacramento, Salt Lake City, St. Louis, Cincinnati, and New York—public health officials collected a gallon of milk each month and sent samples to be measured for radioactive contents. In May 1958, government officials began publicly releasing figures about fallout contents in milk. While the intent of the U.S. Public Health Service and Atomic Energy Commission reports was to minimize the risks of contamination, the monitoring program linked radioactive isotopes to the nation's milk supply in the public's mind, even though vegetables and fruits carried greater concentrations.[32]

Moreover, the emphasis on radioactive milk reflected the long history within the consumer movement that emphasized pure foods, and the involvement of activists who had been part of these earlier campaigns. Historians have well established the importance of scientists like Barry Commoner and D. Evarts Graham to the CNI, but consumer activists constituted an equally important constituency of its membership.[33] Their involvement helped lift the concerns about the effects of strontium-90 on children and infants to the top of the group's agenda. Through their work to screen and publicize the presence of strontium-90 in the nation's milk supply, the Consumers Union (CU) and the Committee for Nuclear Information addressed parents and bewildered consumers alike.

In 1958, the CNI embarked on an ambitious project to better determine children's exposure to radioactive isotopes. The study, proposed by the pediatrician Dr. Alfred Schwartz, sought to determine at what levels strontium-90 accumulated in children's bones and to clarify the range of individual variation in the isotope's absorption. The committee invited parents to send children's baby teeth, accompanied by the child's birthdate and a description of their diet. Between 1959 and 1968, the St. Louis committee gathered thousands of baby teeth from cities as distant as Pittsfield, Massachusetts, and Caldwell, Idaho. Researchers and volunteers organized teeth chronologically and measured their radioactive content in reference to nuclear test schedules, tracing how nuclear testing affected strontium-90 uptake. Consumers Union funded the first analysis of baby tooth data for CNI.[34]

As CNI conducted the baby tooth survey, researchers for CU embarked on their own testing program for milk in forty-eight American cities and two Canadian ones. Their program was five times larger than that of the joint Public Health Service and Atomic Energy Commission project, which tested milk from ten major urban milksheds regularly.[35] The results of the study were published in *Consumer Reports* in 1959. Readers requested reprints, and CU even circulated a short film to "interested readers, group leaders, and schools, PTA's [*sic*], trade unions, church groups and other organizations."[36] By emphasizing the impact of fallout on children's health in familiar forums, CU and CNI brought national attention to a problem marginalized by many congressional leaders.

Antinuclear and consumer activists were not the only ones to illuminate the potential health effects of nuclear testing on the milk supply. In February 1958, Columbia University researchers revealed that rates of strontium-90 deposited in human bones had increased by one-third between June 1956 and June 1957. The study also concluded that strontium-90 aggregated more quickly in the bones of children under the age of four than in adults.[37] The same year, scientists with the National Weather Service and Atomic Energy Commission concluded that more strontium-90 fell in the northerly latitudes. While the scientists' reports did not straightforwardly state it, the areas that experienced the greatest concentration of strontium-90 fallout corresponded with the nation's dairying districts. A knowledgeable reader could infer that as the concentration of strontium-90 increased more rapidly in New York City or Minneapolis, the pasture grasses and alfalfa fields of upstate New York and Minnesota were being disproportionately affected by fallout.[38]

The dairy industry reacted with alarm as the food once lauded as wholesome and healthy became a symbol of the risks of nuclear warfare. When the New York State Health Department announced it would conduct potential spot checks of milk for evidence of radioactive material, one industry journalist expressed concern about the "possible public reaction to ill-considered publicity about strontium."[39] Politicians whose congressional districts included many dairy farms complained about government studies that showed reason for concern.[40] For the dairying industry, lost profits seemed a more real and visible threat than nuclear annihilation.

Dairy companies worked to assure consumers of milk's safety. They asserted, rightly, that cows' bodies filtered out radioactivity. The amount of strontium-90 in cows' ration was seven to eleven times more than was secreted in their milk. Furthermore, dairy scientists noted, cereals and vegetables carried more strontium-90 per unit of calcium than milk.[41] *Consumer Reports* and CNI published the same facts. The magazine counseled readers that it would be "as foolish to stop drinking milk as it would be to refuse an X-ray examination for a broken limb."[42] The CNI reminded readers that "milk is an essential nutrient…and is actually lower in its strontium to calcium ratio than most other foods."[43] But turnip greens laced with radioactive residues inspired far less concern than milk tainted with the same substances. Since Americans believed milk essential to their diet and associated milk with children, they worried more about milk than other foods.

As public concern deepened, dairy companies looked for ways to shore up milk's reputation. Parents sought a way to protect their milk-drinking children. Barring a test ban, which peace activists embraced, concerned citizens set their hopes on a technological fix: a method or machine to remove or reduce milk's strontium-90 content. Screening out strontium-90 proved more difficult than eradicating dirt or disease bacteria. The isotope's chemical makeup matched calcium so closely that early efforts to remove the radioactive substance also eliminated milk's valuable nutrient. By the late 1950s, researchers at the University of Tennessee mastered a process to reduce the isotope's presence. *Consumer Reports* carried news of pilot studies to decontaminate milk in February 1962.[44]

Consumers' desire for a technical solution paralleled earlier efforts to purify milk and cream, such as milk pasteurization or cream neutralization. Although consumers' perception of the source of milk's impurity had changed, they still understood the problem of tainted milk as one that merited further intervention between milk drinkers and the cows that produced it. The turn to a technical fix for fallout-tainted milk also demonstrates that while consumers alarmed by radioactive fallout in milk were wary of contemporary technologies like the atomic bomb, they did not reject technology altogether. Rather, like earlier proponents of chemurgy or artificial insemination, those seeking to reduce fallout in milk believed that science and expertise could be harnessed to better human life. The embrace of technological innovation by readers of *Consumer Reports* meant they had something in common with other environmentalists of the 1960s, such as the readers of the *Whole Earth Catalog*, who saw a place for technological solutions to environmental ills.

Readers of *Consumer Reports,* however, allotted a greater role to the state in guiding technological change than their *Whole Earth Catalog* counterparts. In its coverage of technical means to clean up milk, *Consumer Reports* called on government agencies to take action, voicing dismay that they had not already required dairies to install equipment to reduce fallout loads. Just as consumers had pressed state and local regulations to encourage dairies to pasteurize milk in the early twentieth century, postwar consumer organizations urged government agencies to redefine milk purity and require companies to adopt new tools to uphold it. Nashville citizen Audrey Elder's letter to Senator Estes Kefauver provides an example; she wrote, "If a way has been found to handle this problem in connection with the milk...it should be put into effect as soon as possible. The effects of strontium and the other 'debris' from the tests is so unknown how can we afford to take the chance?"[45] *Whole Earth Catalog* readers, on the other hand, spurned regulatory solutions and favored individual economic actions as a way to foster social and environmental change.[46]

A number of features specific to milk made the *Whole Earth* model of individual responsibility via consumption ill-suited to pure milk. It was impractical for individuals to purchase their own equipment, akin to a home pasteurizer, to eliminate strontium-90 from milk. Nor could individual consumers purchase milk from which fallout residues had been removed at the bottling plant. Even after a method to remove strontium-90 from milk was perfected, few dairy companies adopted it. The industry calculated that decontaminating milk would raise milk prices as much as ten cents a quart and feared that a higher price for milk would ultimately drive down consumption rather than restore consumer confidence. Only in St. Louis, home of CNI and the baby tooth survey, did consumers successfully pressure dairy companies to filter out strontium-90 and sell treated milk at a higher cost.[47] Herein lay the limitations of consumer activism on nuclear fallout. Even as they moved the issue of radioactive milk to the center of political discourse, consumer activists were largely thwarted in changing the material substance of the product.

If consumers were unable to limit the presence of strontium-90 residues in their milk, so were farm families frustrated in efforts to rid milk of the contaminant. Farmers could rid milk of antibiotic residues by altering farm practices. They could overcome cream spoilage by keeping milk cool and the separator clean. But farmers could not prevent the introduction of strontium-90 to their farms through careful sanitation, even though their livelihood depended on assurances of milk's safety.

Milk producers became uneasy as consumers worried about fallout threatened to substitute other foods for milk. Eliminating milk was a choice that few parents took lightly. Most were reluctant to abandon a food considered to be an essential staple of children's diets; yet they wanted to protect their children's health. Their letters to government officials convey a mix of urgency, frustration, and helplessness. Phyllis Smith, of Milwaukee, Wisconsin, wrote to Secretary of Agriculture Orville Freeman that if strontium removal was not begun soon, "it seems our only choice is to substitute calcium pills for milk or perhaps use Similac baby formula."[48] Most parents continued to face a difficult paradox: follow nutritional advice and expose one's child to radioactive particles with undetermined risks, or eliminate an essential source of calcium from children's diet to keep them free of the dangers of fallout.

The dairy industry could not afford a boycott. By 1961, the dairy market picture was bleak. Farm returns in some regions dropped to only half or a quarter of their 1940s levels. Dairy industry executives and many dairy farmers blamed the fear of fallout and heart disease as the key factors driving down milk consumption.[49] Unwilling to alter the nation's security strategy, lawmakers realized the acute impact of fallout fears on dairy farmers. In January 1962, President Kennedy held two press conferences expressly to convince Americans that milk was a healthful, not a hazardous, drink.[50]

By the early 1960s, the dismal economic picture in dairying, combined with interest in a different radioactive contaminant in milk, led to a different approach to curbing radioactive residues. In the early 1960s, focus turned to the isotope iodine-131. The substance had a shorter half-life than strontium-90, at 8.5 days. The element mimicked natural iodine; once present in the body, radioiodine concentrated in the thyroid gland. Chronic exposure to radioactive iodine could cause cancer. Children and infants, it appeared, were especially sensitive to thyroid cancers.[51] The shorter half-life of iodine-131 had some important implications for how dairy farm families, milk drinkers, and government officials would address concerns of fallout-tainted milk.

For years, milk's quality had been predicated on a quick trip from the farm to the consumer's doorstep. The emphasis on getting milk to market quickly, unique to fresh milk, meats, fruits, and vegetables, meant that iodine-131 reached consumers in milk before half of the radioactive particles had decayed. While this insistence on freshness reduced the risk of communicable disease and spoilage, it put milk consumers at greater risk of iodine-131 contamination. But iodine-131's short half-life also presented possibilities. Milk drinkers and farm families theorized that by storing milk—or storing the feed on which cows grazed—they might reduce the iodine-131 content and thereby make milk safe to drink.

Consumers who feared radioactive fallout believed that by purchasing evaporated or dried skim milk, they could limit their children's exposure to iodine-131. Letters sent to the baby tooth survey indicate that some families did alter children's diets to reduce exposure to fallout. In 1964, for instance, one mother reported that her daughter, born at a time of a large atomic test, was unable to tolerate formula but was able to drink homogenized milk. Her pediatrician, she related, "suggested I buy homogenized milk in plastic cartons and freeze them for two weeks. This he said might cut the count in half."[52] While freezing milk as this doctor recommended had little effect on how much strontium-90 milk carried, it would diminish the amount of radioactive iodine to which the drinker was exposed, because the two-week time frame exceeded the half-life of iodine-131.

As concerned consumers changed their purchasing habits, the federal government implemented plans to reduce the risks of fallout contamination. By 1961, the Federal Radiation Council acknowledged that fallout carried risks, even at low levels. As a document published in October 1961 detailed, "The most prudent course is to assume there is no level of radiation exposure below which one can be absolutely certain that harmful effects may not occur to at least a few individuals."[53] Thus, the Federal Radiation Council replaced the term "maximum permissible dose" with three "ranges" of daily intake for strontium-90 and iodine-131. If radiation levels reached the upper levels of range II, the Federal Radiation Council, in cooperation with state and local health authorities, would consider initiating action to reduce the risk of contamination, such as altering dairy cattle feeding practices or diverting milk with high levels of contamination to dairy products that would be stored before sale, like cheese or butter.[54]

The countermeasures government bodies pursued to reduce fallout contamination in milk in the early 1960s involved dairy farm families directly. In the summer of 1962, for instance, public health officials in Minnesota decided to take action to quell levels of iodine-131 when the U.S. Public Health Service revealed that the dose measured 30,500 micromicrocuries per liter. Although the level of iodine-131 was still below range III levels (36,500 micromicrocuries per liter), milk producers and agricultural officials worried that the level would increase when Russians conducted further nuclear tests.[55] Thus, in a meeting held in late June 1962, members of the Twin Cities Milk Producers Association, Minnesota Department of Health, and Minnesota Department of Agriculture proposed that farmers should shift cows off pasture and instead provide them with "aged feed"—grain, roughage, and hay that had been harvested and stored for at least twenty-one days. Aged feed, they believed, would carry less iodine-131, because the element would have partially decayed during storage. Producers who cooperated with the voluntary program would be paid a premium for their milk to compensate for increased costs. In August, as the level of iodine-131 in Minnesota milk reached 34,000 micromicrocuries per liter and the Russian government began a new series of nuclear tests, the aged feeding program went into effect.[56]

While the shift away from pasture grasses could reduce iodine-131 levels, it required dramatic alterations of dairy farm feeding practices. The months during which farm families were being asked to remove their cows from pasture corresponded precisely with the months in which pasture grasses were abundant. Farmers fully recognized the potential risks of fallout, but they bristled at the idea of passing up an economical and nutritious food source, all in the name of reducing the prevalence of an unseen contaminant. Farmers feared that using stored feed would increase feed costs and reduce the amount of milk their cows produced. As one Minnesota farmer informed Secretary of Agriculture Orville Freeman, "Cows that were freshened early '62 will dry up completely, where on the grass they would hold on until they were to be confine [sic] for the winter."[57] Furthermore, the added costs to farmers from shifting from pasturage to stored feed were not distributed equitably. As the Minneapolis Tribune reported, "dairymen who use one of the newest management methods—keeping their cows in the paved barnyard all year and feeding them entirely on hay, silage, and grain" experienced the least trouble from the new recommendations, since they already had a supply of aged feed on hand and adequate facilities to store it.[58] In the end, the shift in feeding practices proved short-lived. By mid-September, state officials recommended that milk producers return to normal feeding practices.[59]

The effects of farmers' efforts in the summer of 1962 to reduce fallout on milk were murky. Since the Minnesota program was voluntary, some of the milk reaching market came from cows who continued to graze on pasture grasses and thereby were exposed to higher levels of iodine-131. Organizers had proposed directing milk from pasture-fed cows to manufactured milk products and using milk from cows on an aged-feed diet for the fluid milk market, but maintaining such distinctions was difficult. By the early 1960s, most Minnesota farmers sent milk to market in tank trucks, in which milk commingled. Hence, while the Minnesota program may have reduced the levels of iodine-131, Minneapolis milk was hardly free of the contaminant.[60]

Furthermore, it was difficult to tell whether Minnesota's aged feeding campaign reduced consumer concerns about iodine-131 or merely stirred up more suspicion about the milk supply. The agricultural officials, health experts, and dairy industry representatives who proposed the aged feeding plan asserted that changing cattle feeding practices would cause less disruption to the milk market than promoting the use of evaporated or powdered milk. But publicity about Minneapolis milk producers' efforts made some consumers wary about the city's milk. Using aged feed instead of pasture grasses as fodder for cattle cast suspicion on the pasture, the source so long identified with milk's purity and wholesomeness. Such dramatic transformations provoked new questions and unexpected conclusions.

In the end, the intense concerns about fallout-contaminated milk came to a near halt when the Partial Test Ban Treaty of September 1963 sent nuclear testing underground. Although worries about radioactive residues in milk receded, a deep concern for milk's safety and perceptions of milk's healthfulness did not. As concerns about radioactive strontium-90 and iodine-131 had intensified, so also had anxieties about another set of chemicals of the post–World War II era: organophosphate insecticides. Experience with fallout interested activists in the effects of pesticides on human health and natural integrity; indeed, some of the same individuals who had publicized the risks of fallout moved on to explore other environmental issues. St. Louis' CNI renamed itself the Committee on Environmental Information and expanded to study a variety of environmental topics—a shift in focus that Rachel Carson herself encouraged to the group in November 1963.[61] Having weathered slow milk sales at the height of concerns about radioactive milk, dairy farmers were acutely aware of the need to protect the public's perception of milk. The reaction to pesticide residues in milk—like antibiotics—reflected farmers' choices about how to balance the need for farm profits and protect the public health at the same time.

Pesticides and Purity: A View from the Farm

By the early 1960s, as problems with antibiotic adulteration waned and efforts to free milk of fallout stalled, FDA food regulators revived their push to eliminate pesticides from the milk supply. The FDA's attention to pesticides in the early 1960s was not entirely new. As early as 1949, enough evidence had accumulated about the aggregative effects of DDT in milk fats that the FDA and USDA jointly issued a zero tolerance for DDT in milk, prohibiting its use for dairy herds. Still, in 1960, 25 percent of the milk samples tested by the Borden Milk Company contained a trace or more of pesticides. In the same year, 2.5 percent of the 936 samples of market milk collected by the FDA contained farm chemicals.[62]

Minimizing pesticide residues proved difficult, for the pathways for pesticide contamination were complex and involved multiple parties. Pesticides were, like antibiotics, a technology that farmers might intentionally introduce to their farm to curtail pesky insects. But pesticides could also drift onto a dairy farm from a neighboring property without the knowledge or authority of the dairy farmer. The chemicals might be introduced to a dairy herd through purchased feeds. Thus, the guilty culprit for contaminating milk with pesticides might be someone who sprayed too much fly spray in the barn where the milk originated, someone in a distant state who sold dieldrin-laced hay to dairy farmers, or someone hired by a local government to spray roadsides for gypsy moths. That pesticide contamination could be introduced from so many sources complicated farmers' ability to eradicate the adulterant. Whereas farm families penalized for selling antibiotic-laden milk could take steps on the farm to remedy the error, those who were situated in areas affected by toxic drift or depended on purchased feed had trouble defending their property from adulterants that might endanger their livelihood and threaten their health.

Dairy farm families rarely appear in the historical accounts of the movement to curtail pesticides. Even the most skillful histories of the movement to control pesticides tend to focus largely on the strategies of environmentalists to curtail pesticide use or organized opposition by the agricultural chemicals industry to these efforts.[63] Environmental groups and consumer organizations who challenged indiscriminate pesticide use rarely recognized the challenges pesticides presented to farmers. Some farmers suffered acute illnesses when they used mislabeled chemicals; others endured income losses when aerial spraying made it impossible for their cows to produce pure milk. Some farm people expressed the same fears as environmental activists about the effects of farm chemicals on the health of their families and their cows. But when environmentalists noted that pesticides poisoned domestic animals, they did so largely to link the concerns about pesticides to human health. Rarely did consumer groups publicize the economic losses poisoned cattle represented to dairy farm families.[64] To the extent that dairy farmers increasingly relied on off-farm inputs and shared the same vulnerability to contamination from farm chemicals as their nonfarming neighbors, concerns about pesticides blurred the line between food consumers and farm producers. This interpretation recovers the unrecognized common concerns faced by food consumers and dairy farm producers in the postwar era, even as it explains the rhetorical positioning and material challenges that prevented most farm families and environmental groups from recognizing these connections.

As early as the 1940s, farm journalists urged farm families to incorporate pesticides with caution. The same farm publications that publicized the beneficial aspects of farm chemicals indicated problems associated with their use. In 1944, for instance, Hoard's Dairyman, the leading national dairying publication, cautioned readers that DDT was "no panacea." In one article, R. N. Annend of the U.S. Bureau of Entomology warned that "much testing remains to be done to determine whether DDT will harm human beings, plants, livestock, other animals, and wildlife."[65] He also told farmers of DDT's long-lasting toxicity.

Praise for DDT's effectiveness, though, outweighed cautionary statements. In 1947, Evert Wallenfeldt, an associate professor of dairy industry at the University of Wisconsin-Madison, told *Hoard's* readers that DDT was a "miracle" for the dairy farmer's fly problem. Wallenfeldt encouraged farm families to spray barns every six to eight weeks, to spray cows every two weeks, and to coat milkhouse walls with DDT frequently.[66] Throughout 1947 and 1948, articles at once referred to "warnings" about DDT and in the next breath told farmers to use DDT for fly control in any form they could get it.[67]

Since unabashed praise of DDT can appear naïve to those of the post–*Silent Spring* era, it is worth remembering why farm families so eagerly embraced it as a fly-killing remedy. To farmers, stable flies and house flies presented more than a pesky nuisance; their presence undermined goals of producing pure and plentiful milk. In the summertime, cows irritated by stable flies became fatigued from stomping their hooves to shoo flies from their legs and bellies. Tired cows produced less milk. House flies could transmit disease from manure to cows' udders or into the milk tank. Due to the role of flies in spreading diseases, milk inspectors who examined dairy farms counted the number of flies they encountered as a measure of a farm's sanitation. Rather than view a liberal coating of DDT as a practice that endangered milk's healthfulness, many farmers interpreted chemical fly control as a sanitation procedure to safeguard milk from communicable diseases. To see chemicals like DDT, not flies, as a cause of disease required a dramatic shift in how farm families understood sanitation, health, and nature.[68]

By 1949, regulators restricted DDT's use on the dairy farm. USDA and FDA officials issued a zero tolerance for DDT in milk, preventing dairy farmers from using DDT as fly spray or to treat their barns. Milk was the only food to receive a zero tolerance.[69] In light of the restrictions, dairying experts encouraged farmers to be more diligent about nonchemical methods to reduce fly populations, such as repairing screens in barns and regularly moving manure from the barnyard. Dairy farmers instituted these practices and also substituted other chemicals, particularly methoxychlor, lindane, and pyretheum, to control flies.[70] In the mid-1950s, the use of DDT alternatives came under question. In 1956, the FDA tightened the rules on methoxychlor, granting it, too, a zero tolerance in milk. What was new in the 1960s, then, was not the awareness of the potential for agricultural chemicals to contaminate milk, but rather the technical capabilities and commitment by regulators to enforce the standards established a decade earlier.

The primary focus of FDA-set tolerances on methoxychlor and DDT was to protect milk consumers from ingesting pesticide residues. But pesticides also posed risks to the farm families applying them and to the animals living on dairy farms. Experiences of chemical contamination deepened farm families' appreciation for consumer protections. In northeastern Wisconsin in 1961, a number of farmers purchased a louse powder manufactured by the Pittsburgh

Figure 5.2 This 1946 photograph of a Saginaw, Michigan, fair depicts both the American appetite for dairy products and the enthusiasm for technological transformation in the immediate postwar era. The barn in the background featured a DDT exhibit. Fairgoers might have even learned of the benefits of DDT while eating an Eskimo Pie! Eskimo Pie Collection, Archives Center, National Museum of American History, Smithsonian Institution.

Plate Glass company called "Corona Cattle Dust." The powder was not labeled as a pesticide, but it contained the chemical endrin, due to a mixing mistake of the manufacturer. When Mr. and Mrs. Robert Muck, who farmed near Shawano, Wisconsin, applied the louse powder to their cows, the cows suffered from nausea, weakness, and lack of physical coordination for weeks thereafter. Three cows of the Mucks' herd died, and nine others suffered permanent injuries.[71] Other farmers lost livestock. Three of Carl Lemirande's cows died after being dusted with the substance. George Lesperance lost three sheep, and Edward Haskins's riding pony died after ingesting endrin from biting the backs of treated cattle. Elmer Hocker, who owned thirteen cows and a small calf, could not sell milk for a month. A few others were removed from the Chicago Grade A market, losing the premium price for their milk.[72]

Upon investigation, the Pittsburgh Plate Glass company admitted the product had been mislabeled and issued a recall. By June 1961, the company had paid a settlement to Mr. Hocker and to the cheese factory to which he sent his milk. Insurance agents visited other affected farms and promised compensation. But in late September 1961, Robert Muck reported to Wisconsin congressman Melvin Laird that he had not yet received payment or further information about a settlement. So far as the record reveals, the Pittsburgh Plate Glass company never paid the Muck family compensation for their lost cows or the personal injuries they suffered.[73] When the farm families affected by endrin contamination insisted on protections from their purchase of a mislabeled chemical, their arguments paralleled those of consumer organizations calling for safeguards. Mislabeled chemicals could harm cows and halt milk markets, putting a dairy farmer out of business. In this sense, consumer protections requiring accurate labeling on pesticide packages used on dairy farms were also producer protections.

Mislabeled chemicals posed great problems, but two other pathways of pesticide contamination—aerial spraying and purchase of contaminated feed—proved more vexing and common predicaments. Already in 1952, Harold Bowles, executive secretary of the Arizona Dairymen's League, expressed alarm about the effects of aerial application of insecticides, especially phosphorous insecticides, on the health of dairy cattle. F. C. Bishopp, the acting chief of the Bureau of Dairy Industry, reassured that though "the phosphorous insecticides such as parathion and tetraethyl pyrophosphate are extremely dangerous poisons," livestock were not in danger so long as precautions and recommendations offered by chemical manufacturers were followed.[74] This advice—that farm chemicals were safe, if applicators applied them according to label directions—became a common refrain from many members of the Agricultural Research Service and the USDA in the 1950s and 1960s.[75]

While historians have paid ample attention to the impact of aerial spraying of pesticides on suburban wildlife and birds, the chemicals' indiscriminate use also had profound effects for farm families and domesticated animals.[76] In Michigan, when aerial sprayers coated a wooded area with DDT for gypsy moth control, the milk from cows on five adjacent dairy farms became so contaminated that residues of DDT appeared eleven weeks after the spraying took place. Louisiana milk inspectors traced heptachlor epoxide residues to six dairies in Avoyelles Parish, Louisiana, that had been accidentally contaminated by planes flying for the federal fire ant control program. 2-4-D sprayed by a township for roadside weed and brush control sickened cows and contaminated milk in Tioga County, New York, while dieldrin used in a county-state grasshopper control program showed up in the butter made by a North Dakota creamery. A rural Minnesota man complained that the city of Roseau, Minnesota, continued to blanket-spray the entire town for mosquito control, despite entomologists' recommendations against the practice. As the crop dusters flew over Roseau's golf courses and the

Roseau River, they poured chemicals over alfalfa fields at the fairgrounds that supplied forage for dairy cattle.[77]

Dairy farmers in the South and in California endured the greatest difficulties with toxic drift.[78] In these regions, dairy farms often operated amid pesticide-dependent cotton farmers and fruit growers and thus faced more frequent exposures to the chemicals. In 1960, W. T. Smith, an extension agent in Booneville, Mississippi, warned of the financial losses facing dairy farmers in his county. "Drift from poisoning cotton can contaminate corn, hay, and pastures. The residues of these insecticides may remain on these crops into the winter months... should tests be run on milk or meat animals our farmers will be out of business."[79] Cotton growers' chemical responses to insects spelled financial disaster for surrounding dairy farmers, because as long as the milk from their cows registered pesticide residues, it was withheld from market.

The experiences of Louis Burrus, who farmed near El Paso, Texas, illustrate the challenges that faced dairy farmers located in cotton-growing districts. In 1957, Mr. Burrus stopped using DDT for any purposes on his farm, but the chemical continued to show up in the milk his cows produced. Louis Burrus restricted his cows' access to DDT in every way he could. He used a DDT-free Indian Head Fly Killer and did not chemically treat his alfalfa crop. He tested the grain he purchased as feed supplements to ensure that it did not carry chemical residues. Although Mr. Burrus was doing all he could to limit his cows' exposure to DDT, he could not insulate his cows from all potential exposures. Cotton fields surrounded Mr. Burrus's farm on the west, south, and east. In 1967, when his neighbors sprayed to control a pink bollworm outbreak, his cows were subject to drift from crop-dusting aircraft. Soon, Mr. Burrus was dumping milk with excessive DDT residues and pleading with his senator, "we desperately need a solution to our problem as we have thus far exhausted every means."[80]

Mississippi's farmers could have related to Mr. Burrus's plight. There, too, drifting pesticides from adjacent cotton farms made it especially difficult to ensure milk's purity. When farm patrons in Okolona, Mississippi, were cut off from market due to excessive pesticide residues, Joe Spinks, the manager of Borden's Okolona plant, asked Mississippi senator Joe Stennis for protection. Explaining the farmers' dilemma, he wrote, "We are not fighting the Food and Drug Administration on their testing program because we feel that it is their duty to protect the consuming public. We do feel however that the producers of this milk should be protected from the un-wanted applications of insecticides on his pastures and feed crops."[81] FDA agents informed Senator Stennis that the federal Food, Drug, and Cosmetic Act provided them with no jurisdiction over aerial sprayers. Mississippi dairying families could hardly hope to garner influence with Mississippi's congressional delegation to control such planes; the U.S. senator from Mississippi, Jamie Whitten, was a staunch defender of chemical agriculture and chaired the appropriations subcommittee on agriculture.[82] Herein lay a major stumbling block for dairy farmers affected by aerial spraying. Representatives from areas with the biggest problems with drifting pesticides and contaminated milk were often reluctant to attack the root of the dairy farmers' problem: indiscriminate spraying by surrounding cotton farmers and orchard growers. But some states, including Arizona and Kansas, did pass state laws to tighten regulations on aerial spraying or to license spray pilots.[83]

To resolve problems stemming from toxic drift, dairy farmers with contaminated pastures or fields purchased cattle fodder. But purchased feed also made farmers susceptible to residues. Henry Herring, of the H and D Dairy near Clint, Texas, was the kind of dairy farmer likely to turn to purchasing feedstuffs from off-the-farm suppliers. By 1967, he determined that it was "nearly impossible" to use home-grown cattle feeds and produce noncontaminated milk. Although Herring had not utilized DDT to dust alfalfa for seven to ten years, so much of the pesticide had accumulated on the soils of his farm that it was present in the grain, silage, cottonseed hulls, and alfalfa used for dairy rations.[84] To keep their cows in production, farmers like Herring would have to turn to feed manufacturers for hay, silage, and grains. When farm families purchased cattle feed, they relied upon the seller's pledge that the hay or silage was free of chemicals restricted in dairying. FDA examinations quickly identified purchased feed as an important source of milk contamination.[85]

Instances of pesticide-laden purchased feed affected dairy farmers nationwide, but California and Arizona dairies testified to the greatest problems. Farmers had few avenues to determine whether residues of disallowed chemicals were present in the hay or silage they purchased. Some California farmers demanded that alfalfa hay be guaranteed by the seller as free of DDT before they would buy it. But in years of drought, when hay was in short supply and farmers most needed it, the hay market tilted toward the seller, not the buyer. Sellers balked at certifying that their product was free of residues. Feed dealers felt that if they guaranteed their product, they would be held liable for residues even if the source of contamination rested with farmers' own hay or the feed a farmer purchased from a different business. To provide clearer information, Arizona dairy farmers successfully pressured state legislators to establish a Public Health Laboratory whose tasks included testing samples of dairy feeds.[86]

Dairy farmers might have articulated their demands for feed guarantees as akin to crusades for fuller labeling and product safety, forging ties to a growing consumer movement. But farm families challenged by residue-laden hay did not make these connections. Instead, they compared the tolerance rules on milk to those governing other agricultural commodities and argued that the zero tolerance for milk unfairly burdened them. As A. C. Pollard, of Hilmar Holsteins in Turlock, California, stated, "The zero tolerance for insecticides in milk is extremely difficult to understand when several parts per million can legally be contained in some foodstuffs for human consumption or in alfalfa hay which can be legally sold to any purchaser."[87] That hay sold for one purpose was held to one tolerance, while the same hay sold for dairying had to be free of residues seemed unjust. Farmers called for the FDA to set pesticide tolerances consistently across agricultural sectors. As California farmer Wesley Sawyer reasoned, if scientific studies necessitated a zero tolerance in milk, "it will of course be necessary that the offending chemicals be banished from the market. I can see no other way in which a dairyman … can possibly stay in business if industry and other segments of agriculture are permitted to continue the uncontrolled use of the materials responsible for said contamination."[88] While Sawyer asked that all agricultural commodities be held to milk's tolerances, others demanded that the zero toler- ance rules governing milk be relaxed. In 1964 and 1965, especially, chemical companies, the Farm Bureau, and the Dairy Industry Committee urged congres- sional representatives and USDA officials to establish pesticide tolerances for milk.[89] In July 1965, a National Research Council committee encouraged stan- dards based on "negligible" or "permissible" residues.[90] Requests for pesticide tolerances in milk signaled the extent to which pesticides had become a part of the living landscape.

Dairy farmers' calls for allowances for pesticide tolerances for milk seemed reasonable to those having difficulty keeping the product free of residues, but proposals to relax regulations on milk met firm opposition. To explain the necessity of zero tolerance provisions, FDA officials reminded audiences of milk's cultural significance and appealed to milk protection's central role in the history of food regulation. Milk and dairy foods deserved more cautious treatment than other agricultural commodities with regard to pesticides, FDA officials reasoned, for these foods constituted nearly a quarter of the average American's diet, a greater proportion than any other crop. More important, milk played a key role in the diet of "the weak, the sick, the young, and the aged"— those judged to be especially susceptible to contamination. The unique role of milk in the American diet, FDA officials insisted, necessitated a greater margin of safety for milk than other food items.[91] Ironically, this message of milk's unique cultural significance, established largely by efforts of the dairy industry to pro- mote the product in the early twentieth century, stifled dairy organizations' efforts to receive tolerances similar to other agricultural commodities in the 1960s.

If milk's status as a food for children and the ill provided the cornerstone of the FDA's defense of zero tolerance, this idea also held sway among consumers. As dairy organizations clamored for indemnities, citizen organizations urged policymakers to uphold tough standards for milk safety.[92] Consumers' concerns about milk's purity drew on the food's long-established place in food regulation but were also ignited by the emerging environmental consciousness. Deliberations about milk's zero tolerance status, timed just two years after the publication of Rachel Carson's *Silent Spring,* shocked citizens engaged by Carson's work. A whole class of schoolchildren from North St. Paul wrote to Secretary of Agriculture Orville Freeman to register their concerns about the proposed tolerance for milk.[93]

When consumers objected to the proposed end to zero tolerance for pesticides in milk, their opposition was as much to the loss of an ideal standard as to any changes detected in the substance of milk itself. Their focus on changes to the standard—not to milk—marked their worries as those of the postwar era, when rising concerns about technological adulteration and the simultaneous development of federal milk standards fundamentally altered the way in which consumers understood milk. Once determined by examining a cow's udder for signs of swelling or counting the number of flies in the milk room, food regulators now relied on bioassay tests to clear suspicion of residues and confirm milk's healthfulness. Consumers consequently interpreted milk as much through federal standards as with their taste buds or through visual or olfactory cues.

Farm families, like consumers, felt the impact of federal policies developed to uphold milk's purity. Even as recognition of technological risks to milk remade federal milk inspection laws, biological hazards continued to shape the work of dairy farming. Farmers worried about a milk shipment being rejected due to the presence of aldrin or dieldrin but also contended with the chomping jaws of alfalfa weevils. They recognized the risk posed by penicillin residues but also worried about the swollen udder of a prize cow. Keeping the goods of nature fit to eat still required harnessing cows' processes of reproduction, feeding a balanced ration, and protecting milk from spoilage. But it also required relinquishing some control over nature to keep milk free of human contaminants. More than ever before, then, achieving milk purity required an agonizingly careful balance of environmental resources, technologies, and human skill.

The establishment of new rules on milk's purity in the postwar era reordered the balance between nature and technology. But it also recalibrated the social order. Rules about purity and pollution, anthropologist Mary Douglas reminds us, are not just about uncleanness, but also place a hierarchy on the order of relationships between parts of society.[94] Hazards posed by fallout and chemical residues, combined with strengthening federal milk inspection, amplified the power of federal scientists and milk inspectors as guardians of milk safety, while minimizing farm families' and consumers' ability to protect milk from contamination. Some milk safety efforts transferred authority to farmers unequally; the incorporation of stored hay in the dairy ration, for instance, displaced the authority of farmers reliant on pasture grazing while rewarding adopters of dry-lot dairying.

The state's extended power over milk safety was never complete, however, for its authority and legitimacy as a protector of purity was tested on many fronts. Scientific uncertainties about fallout or pesticides made farm families and consumers alike question federal milk standards. When new techniques, like antibiotics, proved ineffectual against bacterial hazards, it undermined trust in scientific solutions. Thus, if the general pattern by 1970 was to uphold federal standards of purity and encourage the development of a national milk supply, misgivings about this federally driven system remained. Standards of milk purity may have been codified in federal law, but the question of what constitutes pure milk remained an open question. In the ensuing years, as the trends of concentration and consolidation in the dairy industry continued, these doubts would lead some Americans to envision a dairying economy that provided a greater role to consumers and farm families in defining and producing pure milk.

Epilogue

Since the 1970s, efforts to modernize the dairy farm with technological innovations have proceeded at a rapid pace. Dairy farmers manipulate the bodies of dairy cows and the farm landscape in evermore complex ways. To increase milk yields, some farm people supplement time-tested milk-boosting techniques, such as breeding or feeding for high yields with injections of recombinant bovine growth hormone (rBGH). Other technologies present possibilities unforeseen just decades earlier. For instance, by the 1990s, dairy farmers could not only obtain semen from the world's best bulls to inseminate their cows, but could also pay for sex-selected semen, so that each breeding would yield milk-giving cows and not bull calves that contribute less to the farm income. The technology seeks to eliminate even further the inefficiencies of reproduction that have for so long defied human understanding and slowed dairy farmers' efforts to increase the quantity of milk produced. Such developments extend the long-term efforts to make the nature of the farm more predictable and efficient and to eliminate waste.

As farm families embrace technological efficiencies, consumers' pursuit of the natural has intensified. Although by the 1980s, American consumers were accustomed to eating a host of foods bearing long ingredient labels, milk maintained its status in the popular imagination as a quintessentially "natural" food. In pursuit of purity, some seek milk free of rBGH. Others buy organic milk, believing it to be more healthful and natural than that produced with pesticide-laden feed. A few prefer to purchase raw milk, viewing the unpasteurized food as one stripped of the trappings of modern society. Contemporary concerns about milk quality, like technological developments, are part of a longer history. Suspicion about rBGH in milk, for instance, mirrors concerns of the immediate postwar period, when activists raised the alarm about antibiotics, fallout, and other milk contaminants.

Newfound interest in milk's purity is not simply a reaction to new technologies. It derives just as profoundly from economic and political developments of the post-1970 period. The pursuit of pure milk since the 1970s has taken place in a particular social and economic context, one in which citizens are ambivalent and anxious about a changing rural economy and in which individual consumers—rather than government agencies—are increasingly viewed as the preferred agents of social change. Hence, ideas of milk purity became articulated through an appeal to "the family farm" as much as pastoral nature. Further, the quest for milk purity since the 1980s has been waged through campaigns for product labeling and consumer autonomy, instead of demands for greater government regulation. Placing contemporary interests in organic and raw milk within their own moment, as well as in dialogue with the long twentieth century, foregrounds how the politics of consumer society intersect with the landscape of the dairy farm in distinctive ways.

Technological developments of the late twentieth century bear some responsibility for the dairy case laden with cartons of milk bearing the labels "organic" or "rBGH-free." In the 1980s, researchers developed a way to artificially manufacture bovine somatotropin (BST), a growth hormone that occurs naturally in a cow's pituitary gland. Once scientists identified the gene that coded for the hormone's production, they isolated and incorporated it into bacteria to be reproduced. Then, they injected the synthetically produced hormone into cows' bodies, lengthening cows' lactation periods and increasing the amount of milk the cows yielded.[1] Trials of manufactured BST, also known as rBGH, began in the late 1970s, and U.S. officials approved it for retail sale in 1993. The use of a hormone to stimulate milk production was not a new development. Naturally occurring BST (also known as bovine growth hormone) had long performed this function. What was new to the 1980s was that genetically modified bacteria, outside of cows' bodies, manufactured the hormone.

Promoters of rBGH reiterated rhetorical claims made for many farm technologies that preceded it. They promised the hormone would increase cows' milk-producing capacity without demanding additional labor or feed.[2] While prior technologies to boost milk production altered cows' bodies, bovine biology limited the pace of change. Selective breeding, for instance, could only take place within the framework of cows' gestational cycles. Changing the mix of grains and grasses in the feed trough could boost milk production, but cows' appetites and capacity for digestion restricted some foods from being appropriate cattle feeds. The use of a genetically modified species to generate additional BST, however, seemingly allowed humans to surpass these limitations. With every injection, the human influence on cows' bodies deepened.

If the language used to promote rBGH had antecedents, so did the objections raised against it. Like pesticide or strontium-90 residues, rBGH challenged the perception of milk as a quintessential "natural" food. Consumers who conceived of milk as a product generated by cows' bodies objected to milk that originated partly from a human-instigated chemical injection.[3] Further, those opposed to the hormone claimed it harmed cows' health by elevating levels of mastitis in treated cows. Thus, they argued, rBST intensified antibiotic residue problems, since dairy farmers used more antibiotics to treat mastitis.[4] These claims resonated with animal rights activists and those who identified technologies, rather than disease organisms, as the greatest risks to milk's purity.[5]

Although the debate surrounding rBGH echoed earlier discussions of the appropriate relationship between humans and the environment, the political and economic context of its introduction shaped the content and form of the debate over its use. rBGH arose amid a struggling dairy economy and fears about the disappearing dairy farms. When scientists introduced rBGH, the dairy economy was on the skids, despite the increased productivity of cows. As selective breeding and new feeding regimens boosted milk yields, surplus milk drove prices down, and bills for expensive farm improvements mounted. By 1986, the nation's milk surplus was so massive that the federal government, under the Dairy Termination Program, paid $1.8 billion to dairy farmers who agreed to slaughter their cows and stay out of dairying for five years. Reducing the cow population and thereby milk production, legislators hoped, would lower the costs the federal government paid to maintain dairy price supports.[6] Hence, many farm people greeted news of rBGH not as a technological marvel, but as an absurdity, for dairy farmers were suffering from economic problems resulting from too much milk, not too little.[7]

Rural audiences were not the only ones concerned about the economic strains felt by dairy farmers, for a changing farm economy affected the character of the landscape on the edge of America's cities and weakened the economic sustainability of small towns. In the northeast and mid-Atlantic, suburban roads, houses, and retail establishments swallowed up farmland. Defense installations converted land from agricultural to industrial uses in the West and the South. Dairy farmers were more sensitive to suburban development than any other farm group, because commercial dairying required large, long-term investments for specialized buildings and equipment. As suburbs sprawled, dairying families hesitated to spend money on farm improvements because they believed they might not be able to use them in the future. Farm families on the fringes of the city also feared paying higher taxes to fund suburban roads, schools, and utilities. Hence, metropolitan counties witnessed a steady decline in the number of dairy cows in the 1960s and early 1970s.[8]

Even in states like Vermont, where few counties could be classified as metropolitan, the imprint of the city shaped land use development. Vermont's dairy farms had long been reliant on the city; out-of-state residents, particularly Bostonians and New Yorkers, drink most of the state's milk and have since the early twentieth century. But in the 1960s and 1970s, urbanites and suburbanites increasingly came to Vermont to ski and visit its pastoral landscapes. When these tourists built second homes, Vermont dairy farmers experienced some of the same economic pressures as those on the fringes of American cities.[9] Like nineteenth-century processes of industrialization, processes of suburbanization and urban escape changed country and city alike.

On dairy farms that remained in operation, suburbanization encouraged new land use patterns. Since farm families paid higher land taxes with development, they shifted away from using land extensively for pastures and woodlots and intensified production on cropland. USDA researchers concluded that between 1974 and 1982, pastureland declined by 25 percent on farms in metropolitan counties.[10] Ironically, suburbanites and tourists endangered the very landscapes they idealized—rural pasturelands surrounded by wooded hillsides—by seeking out these landscapes as sites of residence and vacation homes.

In the 1960s, historian Adam Rome argues, the loss of open space inspired urbanites concerned about diminished recreational opportunities and poor flood controls to call for land conservation.[11] By the 1970s and 1980s, though, the effects of the loss of open space on rural residents became of vital interest. In the 1970s, state and local governments introduced differential tax assessment rates for agricultural lands to stem the economic pressures on farmers. Others initiated state land use plans to protect rural lands from development. By the late 1970s and early 1980s, these efforts proved ineffective at maintaining farmland in regions most affected by sprawl and stirred considerable political opposition.[12] Although political efforts to protect farmland from suburban pressures foundered, they reinvigorated appeals to the "family farm" in political rhetoric and forged alliances between farmers and environmentalists. Both these developments would be important in the fight waged against rBGH.

A second factor that made the rBGH opponents' appeals to the family farm resonant for popular audiences at the time of rBGH's adoption was the 1980s farm crisis. In that decade, many farmers suffered from an intense credit crunch. Encouraged to invest in new agricultural technologies to take advantage of international grain markets in the 1970s, farm families found themselves deeply in debt in the 1980s. Although dairy farmers were partially insulated from the farm crisis because milk and dairy foods were not part of the export boom, they did experience a sharp downturn in land values and dairy price supports. Pointing to main streets of small towns tattered by the 1980s farm crisis, farm groups warned of continued decline as rBGH accelerated the pace at which farmers left the business.[13]

Activists' concerns intensified when economists calculated that roughly one-third of the nation's dairy farms were ill-prepared to weather the likely economic effects of rBGH adoption. Economists predicted that highly leveraged medium-sized farms, those most affected by the credit crunch in the 1980s, would suffer most from rBGH, for as declining milk prices drove down land prices, mid-sized farmers would not be able to utilize land as an asset to obtain short-term loans.[14] Groups that drove efforts to restrict the use of rBGH, like Wisconsin's Family Farm Defense Fund, began as organizations to protest the economic conditions facing farmers in the 1980s and viewed the new technology as one that would perpetuate the economic problems of that decade. The fears expressed by these activists fueled successful efforts to pass moratoriums on the use of rBGH in Wisconsin and Minnesota in 1990.[15]

Within this context, opponents to the artificial hormone framed their opposition as a defense of the family farm, as much as a protest about milk's purity. According to political scientist Patricia Strach, the phrase "family farm" only emerged in the immediate post–World War II period to describe the effects of rapid technological modernization on family-sized farms and to distinguish individual farms from collectivized farms in communist states.[16] Opponents to rBGH turned to the phrase, arguing that a stance *against* rBGH was a position *for* the future of the family farm. Importantly, opponents of rBGH used this rhetoric on product labels as much as in the halls of Congress. Ben & Jerry's ice cream, for instance, proclaimed their exclusion of rBGH in milk by emblazoning their packages with the phrase "Save Family Farms." In the wake of failed legislative attempts to preserve agricultural lands and a political climate of neoliberalism, they viewed consumer action, rather than a robust legislative agenda, as the best way to curb the use of the new technology. These contextual factors made the debate over rBGH distinctive from prior discussions about technological innovation.

Placing the family farm at the center of the debate on rBGH was an effective strategy devised by the technology's opponents. The phrase simultaneously addressed rural residents' concerns about the economic viability of farming and tapped into long-standing suburban and urban concerns about the loss of open space in rural areas. But the rhetorical emphasis cast the farm as a site resistant to technological change and reinforced the idea that rural spaces are counterpoint to industrial ones. In fact, by the time of rBGH's introduction, the working landscapes characterized as repositories of traditional values and unchanged nature by the phrase "family farm" had long been hybrids of nature and technology radically transformed by the processes of industrialization in the twentieth century. Since the 1910s and 1920s, small and mid-sized farmers cultivated a precise mix of protein-rich pasture grasses, like alfalfa, to improve cows' efficiency and built silos to encourage steady year-round lactation. After the 1950s, they ridded their barns of pigs, chickens, or goats, orienting labor and investment toward specialized equipment for dairy production. Postwar agriculturalists also moved the locus of expertise for breeding from the bull pen to the farm office, where they pored over breeding records and placed semen orders from artificial insemination firms. In other words, the decision of whether to utilize rBGH did not constitute the most critical moment for distinguishing between industrial, capital-intensive farming and traditional means of raising dairy animals; the technology merely extended and amplified the orientation toward capital-intensive production whose origins lay decades earlier.

What instigated the public debate about rBGH, then, was less the technology itself than the questions it provoked among farm people and public audiences. In the wake of economic dislocation, farm people began to question the capital-intensive, mass-market orientation of agriculture and the burdensome debt loads that accompanied it. Faced with a homogeneous landscape of strip malls and parking lots, suburbanites thought more deeply about the implications of their consumer-driven lifestyle on the countryside and the people who lived there. The debate over rBGH revisited earlier concerns about milk's purity stemming from technological hazards but articulated concerns that the 1980s and 1990s brought to the fore.

The second element that marked the movement against rBGH as a phenomenon of its time was that activists put great emphasis on individual consumers as the agents of social and economic change. Although opponents of rBST worked to slow the technology's progress through government channels, they simultaneously mounted a campaign to make consumers the ultimate decisionmakers in the fate of the new technology. The growing political climate of neoliberalism in the 1980s and 1990s made calls for expanded regulatory authority politically untenable and encouraged opponents of rBGH to showcase consumers, rather than regulators, as agents of change. If consumers could be persuaded to refuse to drink milk from cows treated with the hormone, its critics reasoned, state approval or sanction of the use of rBST would have little effect.

Thus, the central issue became whether milk from cows that were not treated with the artificial hormone could be labeled rBGH-free. Since milk from treated cows bore no obvious visual distinctions from that of nontreated animals, and few consumers knew the immediate origins of their milk, only a label could help consumers to distinguish the practices used in its production. By the end of the twentieth century, food labels served both as a canvas for manufacturers to make claims about their product and also reflected consumer groups' demands for reliable information to inform purchasing decisions—through truth-in-advertising laws, requirements for weights and measures, and ingredient lists. Even consumer-driven activism was in some ways dependent on the state because food regulators decided the pitched battles over the permissible language of food labels.[17]

Monsanto, the leading manufacturer of rBGH, firmly opposed any efforts to label milk that would allow consumers to differentiate between milk derived from treated or nontreated cows. The corporation contended that because milk from treated cows bore no significant nutritional or material distinctions, labeling of milk as free of the technologically derived hormone was false and misleading. Monsanto contended that affixing labels indicating that milk came from treated animals was tantamount to attaching a warning to a product proven safe, while labels indicating a product was rBST-free falsely implied a level of superior wholesomeness.[18]

Opponents of rBGH, however, steadfastly supported efforts to label milk, insisting that consumers had a right to know the contents of their food and the conditions in which their food was raised. The use of a label to tell consumers more about the conditions under which a good was produced was not a new strategy. In the nineteenth century, abolitionists used the "free produce" label to sell goods untainted by slave labor, and Progressive Era consumer activists acted in solidarity with unionized workers through their "white label" campaign.[19] Nor were labels emphasizing product origins new to the dairy industry; after all, Land O'Lakes creameries built a national market in the 1920s by trumpeting the origins of its sweet cream butter on the label. In a political context that placed great value on the individual rights of consumers, rBGH critics articulated their position on labeling as one of protecting consumers' freedom in the market-place.

In February 1994, however, the FDA rejected a mandatory requirement that all milk from rBGH-injected cows be labeled. The agency allowed voluntary labeling to identify milk from nontreated cows as free of rBST, so long as the label put the claim in context, with a phrase such as "No significant difference has been shown between milk derived from rBST-treated and non-rBST-treated cows." Companies who sought to label milk as rBGH-free had to certify and document that their suppliers were indeed eschewing use of the supplemental hormone. In short, the FDA guidelines put the burden of proof on farmers and dairy manufacturers seeking to keep milk free of rBST, rather than on those using the hormone.[20] While the FDA recommendations lacked the force of law, Monsanto's lawyers sought to ensure that the guidelines were strictly interpreted, suing any dairy companies whose labels it believed deviated from the FDA's guidelines.[21]

In the decades following FDA approval in 1993 and the labeling decision of 1994, the hormone neither ushered in the greatest hopes of its proponents nor the greatest fears of its critics. rBST did not increase milk yields as dramatically as its promoters had projected. To produce the levels of milk its promoters had promised, farm people had to manage and monitor the diet and health of cows receiving rBST injections with intense scrutiny, a labor-demanding prospect that was hard to achieve in practice. Further, farm adoption rates were not as widespread as some economists had suggested; even farmers who embraced rBGH often used it selectively on just some of the cows in their herds.[22] In 2002, a USDA survey reported that 22 percent of the nation's dairy cows were treated with artificial growth hormone. By 2007, that number fell to 17 percent.[23]

In the end, the key decisionmakers regarding the use of rBGH were not consumers who refused or embraced milk produced with rBGH, or the individual farm people who enthusiastically adopted or rejected the technology. Rather, the middle sectors of the dairy industry—particularly milk-processing plants, retailers, and regulators—played the largest role in encouraging or curtailing its use. In a dairy industry dominated by large milk processors and supermarket distributors, when a processor refuses to accept milk from cows treated with rBGH, it has important implications on the decisions of its farm suppliers. Farmers who fail to oblige with the milk processor's dictate will have trouble marketing their milk.[24] Many chain retailers and dairy companies decided not to accept milk from rBGH-treated cows, because they feared that even if only a few consumers rejected milk from treated cows, it would drive down profits. By 2008, Walmart, Kroger, and Publix sought out milk without artificial BST, and the major milk bottler Dean Foods shifted away from purchasing milk from farmers who adopted the product.[25]

From the perspective of rBST-backers, the decision by milk processors to reject milk from cows injected with rBST is objectionable, because it disallows dairy farmers themselves to choose whether to use an approved technology. Yet milk companies' influence over farmers' decisions about farm technology is not new. Creamery and milk company fieldmen urged farmers to grow alfalfa, build silos, or buy bulk tank coolers in previous decades. By the late twentieth century, though, the market for milk was so centralized and concentrated that dairy farmers lacked alternative markets if they chose not to follow processors' suggestions. This absence of alternatives made the influence of processors and distributors on farm people's behavior profound.

Although the introduction of rBST was not as revolutionary as originally anticipated, it did set into motion new thinking and action on the farm and in the marketplace. rBST inspired some farmers to adopt technologies to maximize the supplemental hormone's effects, such as state-of-the-art feeding equipment to individualize cows' rations and ease the burdens of management. The artificial hormone motivated other farmers to imagine low-capital alternatives to the technological treadmill rBGH encouraged. For some consumers, rBGH rekindled suspicions about the role of technology on foods like milk, encouraging a new quest for an "all-natural" alternative.

One result of the debate over rBGH was to stimulate interest in organic milk. The use of the term "organic" to describe an agricultural method and kind of food originated in the early 1940s, through the works of Albert Howard and J. I. Rodale. At a time when conventional agriculture emphasized capital-intensive technological improvements to minimize human labor, organic agriculturalists advocated labor-intensive methods that eschewed technology. Advocates of organic farming espoused building the soil with compost, controlling insects with natural predators, and killing weeds with mechanical means, instead of relying on chemical fertilizers, insecticides, and herbicides. Further, rather than sell through centralized food distributors, many organic agriculturalists marketed goods directly to consumers. Proponents of organic agriculture thus cast themselves as an alternative and counterpoint to conventional agriculture's emphasis on chemicals, capital-intensive practices, and centralized markets. By the late 1960s and 1970s, many counterculture activists filled their plates with organic food as a way to make the personal political; disillusioned with the public face of political activism, culinary choices became part of a journey of self-renewal.[26]

While the roots of the organic food movement were established decades earlier, organic milk's popularity boomed in the 1990s as the debate over rBGH unfolded. The growing interest in organic milk transpired at the very moment that the organic label became institutionalized. In 1990, the USDA established national standards to govern organically produced commodities. Implemented in 2002, the standards codified the definition of the term "organic" and required products sold or labeled organic to comply with these rules. Organic dairies, for instance, must eschew rBGH, antibiotics, and feed grown with pesticides for at least three years to meet certification standards. Skeptics of biotechnology turned to the organic label for greater assurances of consistency and food quality than in the rBGH-free label, which varied state to state, based on the outcome of state regulations and legal challenges. Constituting only .3 percent of the nation's milk market in 1999, by 2007 organic milk and cream made up 6 percent of national sales.[27] Federal organic certification standards reinforced a vision of milk's purity that credited technological contaminants as the primary threats to milk safety, because they targeted antibiotics, hormones, and pesticides as substances to be eliminated or avoided.

Organic milk also drew the interest of dairy farm families. In the 1990s, going organic offered a way for mid-sized dairy farmers to opt out of technologically intensive dairy production. Premiums paid for organic milk looked especially promising to prospective producers. In some cases organic milk commanded a price twice that of conventional milk. Most organic producers used lower-cost alternatives to capital-intensive animal-tending regimes, relying on rotational grazing, for instance, instead of mechanized feeding equipment.[28] Still, converting to organic production was not a decision to be taken lightly, because a farm must be managed organically for three years before it is certified and can garner higher organic prices. Organic producers incurred added costs, such as purchasing pesticide-free feed, and their cows produce less milk per cow than those conventional dairies. Despite these high conversion costs, between 2002 and 2007, the number of organic dairy farms rose nearly 80 percent.[29]

In some respects, the extraordinary growth in the popularity of organic milk demonstrated the abiding appeal of all-natural milk among consumers, and a more general acceptance of organic principles among farmers. But organic milk's tremendous growth also owed much to a significant departure from the guiding philosophies of the organic movement. In the 1970s, one of the most revolutionary elements of organic farm proponents was that its provisioners distributed their foods through decentralized, small-scale alternative markets—such as food coops, farmers' markets, and independent health food stores, rather than centralized, corporate-driven supermarkets. In the case of organic milk, however, conventional supermarkets were pivotal.[30] By 2006, Walmart sold more organic milk than any other company.[31] Like members of cooperative butter factories who marketed their product nationally through chain stores in the interwar period, organic producers relied on conventional food distribution channels to invigorate economic and ecological sustainability.

With demand for organic milk outstripping supplies, supermarkets sought large, predictable supplies of organic food. By 2007, three companies—Organic Valley Cooperative, Dean Foods' Horizon, and Aurora Organics Dairy—provided more than three-quarters of the nation's organic milk. Aurora Dairy's milk filled the store-brand bottles for Safeway, Costco, Target, Walmart, and Wild Oats supermarkets. As organic dairy came to be dominated by centralized, corporate firms, some citizens worried that dairy farms and retailers eager to capitalize on demand for organics were not practicing in the spirit of the organic movement and compromised the standards set forth by the USDA.[32]

In the early 2000s, journalists and organic farming organizations exposed contradictions between the perceived organic ideal and the actual circumstances of organic dairy production. On large operations in Colorado and Idaho, thousands of confined dairy cows ate from troughs filled with organically produced grain and were milked three times a day. Milk from these farms flowed to supermarkets backed with the organic label. To critics, the practices seemed counter to the anti-industrial, small-scale, pasture-based model to which the term "organic" hearkened, and they weakened the integrity of the organic label as a whole. Critics also worried that the scale and mass quantities of milk that large organic farms produced would endanger the economic viability of farms operating on a small-scale, pasture-grazing model. Their vision of milk's purity rested not simply on the absence of technological residues in milk but also on factors like animals' quality of life, use of farm-grown feeds, and local control.[33]

In response to this criticism, the USDA Organic Certification program issued new rules to better align the provisions of the certification process with the conception of organic. Issued in 2010 and implemented in 2011, the new USDA regulations forbid cows to be continuously confined, required at least 30 percent of an organic dairy cow's nutrients to come from pasture and for organic dairy cows to graze at least 120 days a year, and demanded that organic farmers submit a pasture management plan as part of certification.[34] The revisions seem to benefit small-scale organic farms whose operators are more experienced with and dependent on pasture grazing, particularly those in the upper Midwest, and to disadvantage the large-scale dairies in the West, which rely more heavily on confinement feeding.

The ultimate effects of this policy change, however, may be difficult to disentangle from the effects of the economic downturn on organic dairying. In 2009, for the first time since the 1990s, organic dairy sales declined. High prices for organic alfalfa and grains have been especially difficult for northeastern organic dairies, whose operators rely more on purchased feed. As the price for conventional milk has fallen, cost-conscious consumers are reluctant to bear the bigger gap between the premium prices of organic and conventional milk. Many who continue to choose organic have cut costs by purchasing store-brand organic milk, which is more likely to come from large-scale organic dairies.[35] The state, by setting federal organic milk marketing standards, and supermarket retailers, by establishing milk prices that favor store brands, play an important role in shaping the alternatives to individual milk purchasers.

Questions raised about the meaning of the word "organic" inspired some consumers to embark on a further search for a more natural milk and a more direct relationship with its origins. In this quest, they have turned to a new-old product: raw milk. Debates about whether to drink raw or pasteurized milk are not wholly new; indeed, physicians and consumers challenged pasteurization as it was first adopted in the Progressive Era. But the contours of the debate over raw milk have shifted. Today's raw milk supporters place great emphasis on entrusting consumers, rather than state actors, with verifying the wholesomeness and purity of food and appeal to the revival of small-scale agriculture through decentralized milk distribution. In so doing, raw milk supporters aim to reject not simply the influence of modern technologies on their milk but also disavow the modern regulatory and distributive systems by which most milk reaches consumers. Although raw milk proponents see the food as a "traditional" one, the ideas and meanings they attach to raw milk reflect a particular historic moment of the 1990s and 2000s.

The push for raw milk departs from the general trend toward pasteurization. Pasteurized milk became the standard fare in America by the 1920s and 1930s as health codes, citing the risk of food-borne illnesses, discouraged the distribution of raw milk. In 1947, Michigan became the first state to ban the sale of raw milk entirely. Other states limited farmers from selling raw milk except directly to consumers on the farm. But as was the case with other laws governing milk safety, a patchwork of state and local laws prevailed well into the twentieth century. Currently, intrastate sales of raw milk are legal in twenty-eight states;

some of these states permit retail sales while others require consumers to buy raw milk directly on the farm site.[36] The sale of raw milk across state lines, however, has been illegal since 1987, when a federal judge ordered the FDA to prohibit interstate shipments.[37]

For its present-day backers, raw milk represents much of what organic foods symbolized in the 1970s: a counterpoint to the highly centralized, corporate-dominated, technologically dependent food system. Raw milk enthusiasts portray the food as natural and untouched by modernity. Writer Nina Planck, for instance, calls raw milk a traditional food, which she contrasts with industrial ones. David Gumpert calls raw milk wild. Sally Fallon Morell describes pasteurization as a *denature*-ing of the product.[38] Like food activists of the late 1960s and early 1970s, today's raw milk enthusiasts view the highly processed foods filling supermarket shelves as symbols of the ills of an industrial society and seek to restore the role of "nature" in their diets.[39] For these proponents, the term "raw" means more than simply unpasteurized. Rather, they associate raw milk with a particular set of practices of food production and distribution. The food's backers associate raw milk with small, local farm operations and link pasteurized milk to large-scale dairy operations. They view raw milk as the product of pasture-grazed cows and connect pasteurized milk to feedlot-fed animals.[40]

While rhetorically effective, the sharp distinction raw milk enthusiasts draw between "traditional" and "industrial" milk misrepresents dairy history. Even seemingly natural and traditional processes, like pasture grazing, have been transformed over the twentieth century. Pastures were not and are not simply plots of native grasses, but carefully managed stands—fertilized, seeded, and fenced to promote the growth of the most succulent and nutrient-rich grasses.[41] By the 1920s and 1930s, for instance, farmers carefully followed advice from modern experts to plant high-protein grasses, like alfalfa, to stimulate milk production. Industrial ideals—maximizing production and minimizing costs—guided interwar farmers planting pasture grasses just as fundamentally as they motivate managers of dairy feedlots today.

Raw milk enthusiasts also celebrate the process by which consumers obtain the food. In many states, raw milk can only be offered for sale at the site of its production. Some raw milk enthusiasts purchase it by entering cow-share agreements; they pay a farmer a fee to house and feed the cow, in exchange for its milk. The intimate set of relationships through which consumers purchase raw milk contrast sharply with the anonymity through which they obtain foods in most retail settings.[42] Raw milk distribution approximates the decentralized, underground, face-to-face model once associated with organic foods.

The decentralized, direct interactions through which raw milk is sold are also attractive to those who produce it. Direct sales promise much higher returns for farmers than they realize from selling milk to processors to be pasteurized or manufactured into other dairy foods. As the number of milk processors contracts, farmers have fewer bidders for their product and less bargaining power. In some markets, farmers must settle for the terms offered by one milk processor.[43] Selling raw milk directly to consumers offers an alternative. One Wisconsin dairyman reports making six times as much for a gallon of raw milk sold directly to consumers than he could obtain from the dairy company.[44] For dairy farmers weary from life on the economic margins, such benefits are compelling. The emergence of the raw milk underground, then, is as much a result of changes in the dairy economy as it is a reflection of contemporary desires for natural foods.

Ironically, the decentralized process by which consumers obtain the food is itself a product of state regulation. The small size of some raw milk dairies, and the direct contact with producers that raw milk drinkers enjoy, derives in no small part from the regulatory restrictions on its sale. In Oregon, for instance, only farms with fewer than three dairy cows can legally sell raw milk, so the small scale of raw milk dairies there reflects less a commitment to traditional practice than compliance with state regulation.[45] The direct connections with producers that many raw milk drinkers value result from the restriction of the food from mass retail settings or interstate shipment.

As raw milk drinkers characterize the food as one free of the trappings of modern life, they also argue that the fight for access to raw milk is about consumers' right to freely make choices about their foods, free from state interference. Raw milk drinkers see the state as an impediment to individual consumer choices, rather than an ally in ensuring milk quality. For many, on-farm interactions and personal feelings of well-being provide more reassurance of the food's safety than evidence of government inspection or scientific claims. Raw milk dealers who have been prosecuted for circumventing public health laws view such actions as part of an orchestrated government plot to stifle small farms and halt the rights of consumers. Some raw milk defenders intimate that FDA inspectors are acting to defend agribusiness from competition, while others argue that actions taken against raw milk mark an over-expansion of state regulation over food.[46] As farmer Joel Salatin, whose views have received widespread exposure through Michael Pollan's *Omnivore's Dilemma* and the documentary *Food, Inc.*, writes, "When the public no longer trusts its public servants, people begin taking charge of their own health and welfare. And that is exactly what is driving the local heritage food movement."[47]

The libertarian-leaning, antiauthoritarian views of raw milk drinkers reflect the changing political culture of American life after the 1970s. In the 1970s, political conservatives challenged the popular consensus backing consumer protection laws, casting state protections as a threat to individual consumer freedoms. Rather than argue that the state had a legitimate role to ensure the rights and protect consumers as a body, conservatives emphasized the rights of each individual consumer to act unbridled in the marketplace. This emphasis on individual agency and autonomy also flourished among countercultural communities who emphasized individual responsibility and do-it-yourself solutions to broader environmental and social problems. Even as raw milk proponents celebrate their drink as antimodern, they also articulate views consistent with those who sought to reconcile technology and nature in the 1970s and 1980s.[48]

Contemporary raw milk enthusiasts' ideas break from past precedents on milk regulation in two ways. Although raw milk drinkers see themselves as part of a consumer movement, historically, most consumer activism surrounding milk has called for government intervention on behalf of milk drinkers, not freedom from it. In the Progressive Era, physicians and parents concerned about milk quality pushed for the creation of municipal ordinances to improve milk's sanitary qualities. By the 1950s and 1960s, even as they criticized the federal government's role in nuclear testing and indiscriminate pesticide use, consumer activists called for Congress to enact tougher standards to improve milk's purity and sought to toughen FDA enforcement. Citizens saw these public health interventions that restricted individual liberties to be justified through the constitutional principle that grants the state police power to promote the public's general health and well-being. By contrast, raw food activists regard state health inspectors, and the FDA in particular, as an oppressive force that stifles individual consumers' rights to access healthy food.

Second, raw milk enthusiasts depart from the position, widely established in the Progressive Era, that many of the bacteria milk carries pose potential health hazards. Many of the illnesses traced to milk in the Progressive Era—septic sore throat, typhoid, and scarlet fever—have subsided with safeguards to sanitize the water supply, control veterinary diseases, and improve dairy workers' health. However, public health officials still worry about the risks posed by food-borne illnesses such as *Campylobacter jejuni, Salmonella dublin,* and *E. coli* 0157:H7.[49] Whereas public health officials stress raw milk's role in communicating food-borne pathogens, raw milk supporters characterize the food as an elixir of health. They focus on how it helps remediate symptoms of chronic diseases, such as arthritis, eczema, asthma, and childhood behavioral disorders.[50] Although raw milk has long had promoters, the emphasis on chronic diseases and immune disorders, and their lack of concern about food-borne illnesses, distinguish today's raw milk movement from its predecessors.

Ultimately, today, as in the early twentieth century, regulators' decisions about whether and in what ways to allow the sale of raw milk turn on economic considerations. In the early twentieth century, when raw milk commanded a greater proportion of the milk sold, certified raw milk dairies were subject to more regular inspection by medical milk commissions and had to meet lower bacterial standards than pasteurized milk. At that time, the high cost of on-farm inspection is what drove many municipalities to pass pasteurization ordinances. Today, the cost of inspection of raw milk often presents the biggest hurdle to the food's legalization, because states bear the expense of ensuring that the unpasteurized milk offered at the farm is free of bacterial contaminants.[51] Legalizing raw milk requires the maintenance of sufficient staffing to conduct farm inspections and test milk samples in food laboratories. Maintaining a robust milk inspection system is costly. In cash-strapped times, devoting state resources to a food that reaches only 1 to 3 percent of milk drinkers seems to outweigh the food's benefits.

Conflicts over milk labeled raw, organic, or rBGH-free suggest that competing cultural ideals and political contests are at the heart of Americans' challenges to find a sustainable balance with nature. Consumers, producers, and health officials present diverse ideas about the risks of milk and who should bear the responsibility for food safety. But such cultural and political discussions always take place in a biological context. As policymakers and citizens deliberate over the language to print on milk labels and the meaning of organic, the nature of the farmstead continues to undergo change, presenting new challenges to health officials and farm families charged with protecting milk's purity.

For instance, in the past fifteen years, health officers have begun to track the re-emergence of some of the bacteria whose presence sparked pasteurization ordinances. In the late 1990s, state health officials began to detect isolated cases of bovine tuberculosis—the same disease that prompted Progressive Era battles over tuberculin testing—in livestock herds. The resurgence of bovine tuberculosis was surprising, for federal-state testing of cattle on regular intervals had largely eradicated the illness. So few cases were detected that by the early 1990s some states, such as California, abandoned testing. Less than five years later, though, bovine tuberculosis appeared in California cattle, and by 2010, cases had also turned up in states throughout the country.[52]

The recent outbreak of bovine tuberculosis offers both an opportunity to examine contemporary developments in relationship to the past and also challenges present-day assumptions about milk purity, nature, and consumer culture. First, at a time when questions of milk's purity among consumers had come to focus so centrally on the risks posed by technological contaminants, the resurgence of bovine tuberculosis presents consumers with a reminder of the ways that turn-of-the-twentieth-century milk drinkers conceived of milk's hazards. Although the widespread adoption of pasteurization and the ready availability of antibiotics to battle the illness means that few consumers could be afflicted, bovine tuberculosis still poses a potential risk to the health of dairy cattle and reduces the amount of milk that cows can produce. The resurgence of bovine tuberculosis challenged notions of natural purity not simply because it reminded consumers of bacterial hazards, but also because of the vector of its introduction to dairy herds. In Michigan and Minnesota, wildlife, particularly deer populations, spread the disease. Ironically, at the very moment that consumers came to credit nature for milk's purity, the rewilding of the rural landscape reintroduced the germs of bovine tuberculosis to a select few dairy herds.

In this respect, the spread of bovine tuberculosis exposes the legacy not simply of Progressive Era health laws, but the changing fate of farming since the 1960s. When farm families left the business in the postwar era, some allowed once-grazed pasturelands to reforest. Others sold their farms to developers to be converted into residential lots or recreational lands. Road construction that allowed easy access to homeowners and nature enthusiasts crowded out natural predators and stimulated the development of edge species that fed deer populations. Exacerbating the trend, members of rural hunt clubs baited deer populations during hunting season. Close contact between animals at deer baiting sites, like the proximity of cows enclosed in dairy barns, facilitated the spread of tuberculosis.[53] The shift from farmland to recreational land after World War II carried broad implications, then, not simply for displaced farm families or the suburbanites and recreationists who replaced them, but also as it remade ecological relationships and thereby the livelihoods of those who continued to farm.

A third element to which recent cases of bovine tuberculosis draw attention is that broader biological and technological changes transcend the boundaries erected between varieties of dairy in the marketplace. Just as southern dairy farm families in the 1950s found it difficult to avoid toxic drift, and organic farmers today fear that genetically modified alfalfa will make it impossible to market milk as organic, the resurgent spread of bovine tuberculosis demonstrates the ecological and economic connections that link farms to one another. Whether they used rBGH or not, sold milk for local sale or to be churned into butter, grazed or fed animals in confinement, all farmers felt the potential risks of a bacterial hazard and the tightening measures to control the disease. Farmers in regions with active cases of bovine tuberculosis, for instance, found it more difficult to move cattle across state lines. By reminding farm people and consumers of a common fate, the disease called into question the divisions between sectors of the dairy industry highlighted by fights over rBGH or organic agriculture.

Citizens currently express considerable optimism about the ability of consumers as eaters to remake farm practices. Hoping to improve the economic viability and environmental vitality of the countryside, some flock to farmers' markets. Others sign up for community-supported agriculture shares. Still others fill their grocery cart with fair-trade coffee or local foods. These citizen-shoppers are inspired by catchy slogans that invest food choices with great significance: Eat Your View! Vote with Your Fork! Individual eaters see market transactions as a way to support local agricultural economies, to improve water and soil quality, and to limit farmworkers' exposure to carcinogenic substances. It is not altogether surprising that such efforts have yielded interest. The act of purchasing a delectable artisanal cheese or an heirloom squash provides an immediate tangible reward, as well as a way to demonstrate a sustained commitment to improving environmental quality.

In many respects, this history of milk and dairy farming in the twentieth century offers reasons to see these developments in a positive light, for it reveals tight connections between the fate of the dairy farm and consumer practice. After all, were it not for new ways of thinking about the consumption of milk in the early twentieth century, dairy farmers might not have been able to peddle fresh milk as a staple food. Only when mothers became doubtful about their capacity to breastfeed their children did they turn to cows' milk as an alternative. Over time, changing consumer practice led dairy farmers to do things once unthinkable, like conceive of skim milk as a product to be directed to consumer markets rather than used for animals on the farm. And as Americans spurned butter for lower-fat or cheaper alternatives, dairy farmers adjusted by remodeling the barnyard so they could sell milk to Grade A processors. Each of these historical examples about milk underscores a premise central to contemporary efforts to reform the food system: that the relationship between consumers and farm families is tightly linked and that consumer demands can prompt reflexive thinking and changes in practice by food producers.

But the lessons the history of milk teaches are not all so simple. The history of dairying also reveals that the power of consumers is always mediated through the intermediate sectors of the commodity chain, that is, by health officials, retailers, food processors, nutritionists, and advertisers. In the case of butter, it was fieldmen working for dairy cooperatives and chain store retailers, not individual consumers, who articulated the value of sweet cream butter and convinced farm people to adopt new techniques to make a more uniform product. By the 1940s, rationing programs limited consumers' access to butter, forcing them to either spurn the food or obtain it on the black market. In the 1960s, those who sought to obtain milk with reduced levels of radioactive contaminants were hindered from doing so by dairy companies and health officials who doubted that most consumers would be willing to pay for decontaminated milk. In the case of rBGH, supermarket retailers and food processors ultimately limited the use of rBGH by dairy farmers by refusing to purchase milk from animals treated with the synthetic hormone. Hence, even in a political moment seemingly friendly to consumers' power in the marketplace, the political and economic structures of contemporary society frame the kinds of choices consumers have in the marketplace, and in so doing, the environmental consequences of such choices.

Milk's history reminds us that present-day consumers' desire to see individual food purchases as a means to engage broader political issues is not a new phenomenon. Butter boycotters flexed their buying power to criticize the high cost of living in the 1910s. Ice cream eaters pressed the FDA for more transparency on the contents of the food in the 1950s. Postwar parents questioned the effects of farm chemicals on milk's purity. Over the course of the twentieth century, however, consumers advocating change in the dairy industry have come to place greater emphasis on the rights of individuals to make choices in the marketplace, rather than on improving access and setting standards for all consumers to obtain pure and healthful food. Pure milk reformers at the turn of the twentieth century wanted to improve the diet of their own children, but also worried about the fate of poor children who suffered from unsafe milk. As a staple food, conceptualized by nutritionists as essential to the diet of all, consumers urged shared responsibility for food safety. But as the American diet has become more diverse, the trust in nutritional science less absolute, and food-borne illnesses less common, consumers' movements are less likely to push for a universal standard of milk purity than to defend their own, individual access to a particular kind of dairy food. Hence, despite the potential inherent in the collective efforts of individual consumers to affect farm practice, present-day efforts are, in some ways, more limited than their predecessors. Transforming one's own food purchasing practices is a start, but remaking the agricultural system to be economically and ecologically sustainable and for its bounty to be within the reach of all requires more robust political action.

Rhetorically, present-day battles over rBGH or organic food tend to reiterate a distinction between nature and technology, for instance, characterizing milk from cows treated with artificial BGH or housed on feed-lots as products of technology and non-rBGH or grazing animals as products of nature. The history of milk and dairy farming illustrates that whether on an organic or a conventional farm, large or small, in the distant past or more modern times, turning cows' milk into a human food requires a careful mix of technological and natural forces. At the turn of the twentieth century, each of the fodder crops that made up a cow's ration—whether hay, silage, soybeans, cottonseed meal, or other concentrates—had to be planted, cultivated, harvested, and hoisted into farm buildings to last the winter. Even before mechanization, milking required equipment—such as pails and cans to carry and store milk, and a pump to capture spring water or groundwater to use to cool it. Horses drew loads of ice to facilitate refrigeration and the carts that delivered milk to the processing plant or rail station. While processes of animal feeding, milking, cooling, and transportation became increasingly mechanized, capital-intensive, and large-scale over the course of the twentieth century, they still required a blend of human intervention and natural action. Products of the dairy farm historically, and in the present, are hybrids of human technologies and nonhuman environments.

Furthermore, despite our proclivity to think of the threats posed to food safety as either the result of natural phenomena or technological mistakes, many of the most challenging problems with milk's purity and dairy production have stemmed from the interplay of natural and technological forces. For instance, butter factories encountered the biggest problems with cream spoilage after the introduction of the cream separator and the automobile. Natural spoilage tainted milk, but unclean separator parts that harbored these bacteria intensified off-flavors. Similarly, solutions to improve the safety of dairy foods and the productivity of the dairy farm have incorporated natural and technological variables. Artificial insemination boosted production by using new mechanisms to introduce bull's semen to the cow's uterus, but still had to be performed at the appropriate moment in a cow's estrous cycle and required the cow's work in carrying the nascent calf to term. Such examples remind us that neither nature nor technology is inherently risky. Neither nature nor technology is inherently good. Rather, both the problems of impure milk and the solutions to farm productivity have required a muddled mix of nature and technology.

The murky and ever-changing lines between nature and technology have inspired many of the changes to milk and the dairy farm over the course of the twentieth century. Among farm people and agricultural scientists, the persistent influence of natural cycles and unpredictable outcomes on the farm inspires a turn toward technologies to control variability and make farm production more efficient. Among some consumers, the pervasive influence of technology on nature raises discomfort about compromising the natural purity of foods. Only by recognizing this muddled mix of environmental forces and human choices inherent in a glass of milk, an ice cream sundae, or a whey-protein shake can we come to terms with the ways in which humans have manipulated and transformed nature in the twentieth century.

The most promising feature of recent interest in food and agriculture is that it encourages eaters to think of the act of eating as one with consequences beyond the realm of the individual. Decisions about food shape not simply the contours of one's waistline or one's score on a cholesterol test, but also the health of the broader ecological and social fabric of which we are part. Food constitutes community. What eaters put on the plate or pour in the glass embeds them within social and economic relationships, engages them in political contexts, and ties them to ecological processes. By explaining how the natural and social networks inherent in a nibble of cheese or a swig of milk have changed over time, this book hopes to inspire greater consideration of these consequences and the communities that sustain us.

NOTES

Introduction

1. Edna Irwin to Betty Furness, November 20, 1967, box 4644, Record Group 16, Records of the Office of the Secretary of Agriculture, General Correspondence, 1906–1976. National Archives and Records Administration, College Park, Md. (hereafter RG 16, SOA, GC, 1906–76, NARA).

2. Nicolaas Mink, "It Begins in the Belly," *Environmental History* 14 (April 2009): 313.

3. Douglas Cazaux Sackman, *Orange Empire: California and the Fruits of Eden* (Berkeley: University of California Press, 2005), 82–83; Marian Burros, "'Natural' Food: Telling the Real from the Artificial," *New York Times*, August 18, 1982.

4. Eric Lampard, *Rise of the Dairy Industry in Wisconsin* (Madison: State Historical Society of Wisconsin, 1963); E. Melanie Dupuis, *Nature's Perfect Food: How Milk Became America's Drink* (New York: New York University Press, 2002); Anne Mendelson, *Milk: The Surprising Story of Milk Through the Ages* (New York: Knopf, 2008); Peter Atkins, *Liquid Materialities* (Surrey: Ashgate, 2010); Andrea Wiley, *Re-Imagining Milk* (New York: Routledge, 2011); Deborah Valenze, *Milk: A Local and Global History* (New Haven: Yale University Press, 2011); Daniel Block, "Development of Regional Institutions in Agriculture: The Chicago Milk Marketing Order" (Ph.D. diss., University of California, Los Angeles, 1997); Jess Gilbert and Kevin Wehr, "Dairy Industrialization in the First Place: Urbanization, Immigration, and Political Economy in Los Angeles County, 1920–1970," *Rural Sociology* 68 (December 2003): 467–490; Barbara Orland, "Turbo-Cows: Producing a Competitive Animal in the Nineteenth and Early Twentieth Centuries," in *Industrializing Organisms: Introducing Evolutionary History*, ed. Susan Schrepfer and Philip Scranton, 167–190 (New York: Routledge, 2004); Susanne Freidberg, *Fresh: A Perishable History* (Cambridge: Belknap Press, 2009), 197–234; Shane Hamilton, *Trucking Country: The Road to America's Wal-Mart Economy* (Princeton: Princeton University Press, 2008), 25–34, 163–186; Daniel Block, "Public Health, Cooperatives, Local Regulation, and the Development of Modern Milk Policy," *Journal of Historical Geography* 35 (2009): 128–153; Manfred Waserman, "Henry L. Coit and the Certified Milk Movement in the Development of Modern Pediatrics," *Bulletin of the History of Medicine* (July–August 1972): 359–390; Judith Walzer Leavitt, *The Healthiest City: Milwaukee and the Politics of Health Reform* (Princeton: Princeton University Press, 1982), 156–189; Rima Apple, *Mothers and Medicine: A Social History of Infant Feeding, 1890–1950* (Madison: University of Wisconsin Press, 1987); Richard Meckel, *Save the Babies: American Public Health Reform and the Prevention of Infant Mortality, 1850–1929* (Baltimore: Johns Hopkins University Press, 1990); Alisa Klaus, *Every Child a Lion: The Origins of Maternal and Infant Policy in the United States and*

France, 1890–1920 (Ithaca: Cornell University Press, 1993); Jacqueline Wolf, *Public Health and the Decline of Breastfeeding in the 19th and 20th Centuries* (Columbus: Ohio State University Press, 2001).

5. Tom McCarthy describes this theme in the literature as the "big connections" school, "The Black Box in the Garden: Consumers and the Environment," in *A Companion to American Environmental History*, ed. Douglas Sackman, 312–331 (Malden, Mass.: Wiley-Blackwell, 2010). Examples include William Cronon, *Nature's Metropolis: Chicago and the Great West* (New York: Norton, 1995); Jennifer Price, *Flight Maps: Adventures with Nature in Modern America* (New York: Basic, 1999); Richard Tucker, *Insatiable Appetite: The United States and the Ecological Degradation of the Tropical World* (Berkeley: University of California Press, 2000); Kathryn Morse, *The Nature of Gold: An Environmental History of the Klondike Gold Rush* (Seattle: University of Washington Press, 2003); John Soluri, *Banana Cultures: Agriculture, Consumption and Environmental Change in Honduras and the United States* (Austin: University of Texas Press, 2005).

6. For farm as counterpoint to urban industrialism, see Christopher Sellers, *Hazards of the Job: From Industrial Disease to Environmental Health Science* (Chapel Hill: University of North Carolina Press, 1999), 3; Raymond Williams, *The Country and the City* (Oxford: Oxford University Press, 1975); for environmentalists' approach to rural work, see Richard White, "Are You an Environmentalist or Do You Work for a Living? Work and Nature," in *Uncommon Ground: Rethinking the Human Place in Nature*, ed. William Cronon, 171–185 (New York: Norton, 1995), 180–182.

7. Donald Worster, *Dust Bowl: The Southern Plains in the 1930s* (Oxford: Oxford University Press, 1979); John Opie, *Ogallala: Water for a Dry Land*, 2nd ed. (Lincoln: University of Nebraska, 2000).

8. Jeffrey K. Stine and Joel A. Tarr, "At the Intersection of Histories: Technology and the Environment," *Technology and Culture* 39 (October 1998): 601–640; Martin Reuss and Stephen H. Cutcliffe, *The Illusory Boundary: Environment and Technology in History* (Charlottesville: University of Virginia Press, 2010).

9. Richard White, *The Organic Machine: The Remaking of the Columbia River* (New York: Hill and Wang, 1995).

10. Cronon, *Nature's Metropolis*; Martin Melosi, "The Place of the City in Environmental History," *Environmental History Review* 17 (Spring 1993): 1–23; Martin Melosi, ed., *Effluent America: Cities, Industry, Energy, and Environment* (Pittsburgh: University of Pittsburgh Press, 2001); Craig Colton, *An Unnatural Metropolis: Wrestling New Orleans from Nature* (Baton Rouge: Louisiana State University Press, 2005); William Deverell and Greg Hise, *Land of Sunshine: An Environmental History of Metropolitan Los Angeles* (Pittsburgh: University of Pittsburgh Press, 2005); Michael Rawson, *Eden on the Charles: The Making of Boston* (Cambridge: Harvard University Press, 2010).

11. Andrew Hurley, *Environmental Inequalities: Class, Race, and Industrial Pollution in Gary, Indiana, 1945–1980* (Chapel Hill: University of North Carolina Press, 1999); Sellers, *Hazards of the Job*; Michelle Murphy, *Sick Building Syndrome and the Problem of Uncertainty: Environmental Politics, Technoscience, and Women Workers* (Durham: Duke University Press, 2006).

12. Deborah Fitzgerald, *The Business of Breeding: Hybrid Corn in Illinois, 1890–1940* (Ithaca: Cornell University Press, 1990); Jack Kloppenberg, *First the Seed: The Political Economy of Plant Biotechnology* (Cambridge: Cambridge University Press, 1998); Mark Fiege, *Irrigated Eden: The Making of an Agricultural Landscape in the American West* (Seattle: University of Washington Press, 2000); William Boyd, "Making Meat: Science, Technology, and American Poultry Production," *Technology and Culture* 42 (October 2001): 631–664; Schrepfer and Scranton, *Industrializing Organisms*; Ann Norton Greene, *Horses at Work: Harnessing Power in Industrial America* (Cambridge: Harvard University Press, 2008); J. L. Anderson, *Industrializing the Corn Belt: Agriculture, Technology, and Environment, 1945–1972* (DeKalb: Northern Illinois University Press, 2009).

13. Edmund Russell, "The Garden in the Machine: Toward an Evolutionary History of Technology," in *Industrializing Organisms*, 1.

14. McCarthy, "The Black Box in the Garden," 305; Tom McCarthy, "Follow the Buyer," *Environmental History* 10 (January 2005).

15. See Sidney Mintz, *Sweetness and Power: The Place of Sugar in Modern History* (New York: Penguin, 1985); Cronon, *Nature's Metropolis*; Tucker, *Insatiable Appetite*; Adam Rome, *Bulldozer in the Countryside: Suburban Sprawl and the Rise of Modern Environmentalism* (New York: Cambridge University Press, 2001); Morse, *The Nature of Gold*, 138–165; Douglas Sackman, "Putting Gender on the Table: Food and the Family Life of Nature," in *Seeing Nature through Gender*, ed. Virginia Scharff, 169–193 (Lawrence: University of Kansas Press, 2003); Nancy Shoemaker, "Whale Meat in U.S. History," *Environmental History* 10 (April 2005): 269–294; Sackman, *Orange Empire*; Soluri, *Banana Cultures*; Ann Vilesis, *Kitchen Literacy: How We Lost Knowledge of Where Our Food Comes From and Why We Need to Get It Back* (Washington, D.C.: Island Press, 2008); Freidberg, *Fresh*; Robert Chester III and Nicolaas Mink, "Having Our Cake and Eating It Too: Food's Place in Environmental History—A Forum," *Environmental History* 14 (April 2009): 309–344; Douglas Sackman, "Food," in *A Companion to American Environmental History*, 529–550.

16. For twentieth-century consumer politics, see Lizabeth Cohen, *A Consumer's Republic: The Politics of Mass Consumption in Postwar America* (New York: Knopf, 2004); Meg Jacobs, *Pocketbook Politics: Economic Citizenship in Twentieth-Century America* (Princeton: Princeton University Press, 2005); Lawrence Glickman, *Buying Power: A History of Consumer Activism in America* (Chicago: University of Chicago Press, 2009); Tracey Deutsch, *Building a Housewife's Paradise: Gender, Politics, and American Grocery Stores in the Twentieth Century* (Chapel Hill: University of North Carolina Press, 2010).

17. Ruth Oldenziel, Adri Albert de la Bruhèze, and Onno de Wit, "Europe's Mediation Junction: Technology and Consumer Society in the 20th Century," *History and Technology* 21 (March 2005): 111–115.

18. For bodily knowledge in environmental history, see Christopher Sellers, "Thoreau's Body: Towards an Embodied Environmental History," *Environmental History* 4 (October 1999): 486–514; Conevery Bolton Valencius, *The Health of the Country: How American Settlers Understood Themselves and Their Land* (New York: Basic, 2002); Susan D. Jones, "Body and Place," *Environmental History* 10 (January 2005): 47–49; Gregg Mitman, "In Search of Health: Landscape and Disease in American Environmental History," *Environmental History* 10 (April 2005): 184–210; Gregg Mitman, Michelle Murphy, and Christopher Sellers, "Landscapes of Exposure: Knowledge and Illness in Modern Environments," *Osiris* 19 (2004) 1–17; Linda Nash, *Inescapable Ecologies: A History of Environment, Disease, and Knowledge* (Berkeley: University of California Press, 2006); Gregg Mitman, *Breathing Space: How Allergies Shape Our Lives and Landscape* (New Haven: Yale University Press, 2007); Nancy Langston, *Toxic Bodies: Hormone Disruptors and the Legacy of DES* (New Haven: Yale University Press, 2010); Neil Maher, "Body Counts: Tracking the Body through Environmental History," in *A Companion to American Environmental History*, 163–179.

19. Michael Pollan, *The Omnivore's Dilemma: A Natural History of Four Meals* (New York: Penguin, 2007); Barbara Kingsolver, *Animal, Vegetable, Miracle: A Year of Food Life* (New York: Harper Perennial, 2008).

Chapter 1

1. Editorial, "The Problem of Milk Preservation," *Pediatrics* 16, no. 9 (September 1904): 562.

2. Jacqueline Wolf, *Don't Kill Your Baby: Public Health and the Decline of Breastfeeding in the Nineteenth and Twentieth Centuries* (Columbus: Ohio State University Press, 2001), 41; for

the variety of bacteria, see Horatio Newman Parker, *City Milk Supply* (New York: McGraw-Hill, 1917), 16–17.

3. S. Josephine Baker, *Fighting for Life* (New York: Macmillan, 1939), 108.

4. Irving Fischer, "Infant Mortality and Social Welfare," *Pediatrics* 22 (January 1910): 77–78. Psychologist G. Stanley Hall even included milk reform within his planned "Department of Eugenics." Martin Pernick, "Eugenics and Public Health in American History," *American Journal of Public Health* 87 (November 1987): 1768.

5. Meg Jacobs, *Pocketbook Politics: Economic Citizenship in Twentieth-Century America* (Princeton: Princeton University Press, 2005), 4; Charles McGovern, *Sold American: Consumption and Citizenship, 1890–1945* (Chapel Hill: University of North Carolina Press, 2006), 14–17; Lawrence Glickman, *Buying Power: A History of Consumer Activism in America* (Chicago: University of Chicago Press, 2009), 157–160.

6. Suellen Hoy, *Chasing Dirt: The American Pursuit of Cleanliness* (New York: Oxford University Press, 1996), 81–85; Maureen Flanagan, *America Reformed: Progressives and Progressivisms, 1890s–1920s* (New York: Oxford University Press, 2007), 42–48; Maureen Flanagan, *Seeing with Their Hearts: Chicago Women and the Vision of the Good City, 1871–1933* (Princeton: Princeton University Press, 2002); Sarah Deutsch, *Women and the City* (New York: Oxford University Press, 2000); Adam Rome, "'Political Hermaphrodites': Gender and Environmental Reform in Progressive America," *Environmental History* 11 (July 2006): 442.

7. Rima Apple, *Mothers and Medicine: A Social History of Infant Feeding, 1890–1950* (Madison: University of Wisconsin Press, 1987), 169, 173–175; Harvey Levenstein, "'Best for Babies' or 'Preventable Infanticide': The Controversy over Artificial Feeding of Infants, 1880–1920," *Journal of American History* 70 (June 1983): 80; Alisa Klaus, *Every Child a Lion: The Origins of Maternal and Infant Health Policy in the United States and France, 1890–1920* (Ithaca: Cornell University Press, 1993), 84–89; John Duffy, *The Sanitarians: A History of American Public Health* (Urbana: University of Illinois Press, 1990), 198; Judith Walzer Leavitt, *The Healthiest City: Milwaukee and the Politics of Health Reform* (Princeton: Princeton University Press, 1982), 156–189; on the rise of expertise more generally in this period, see Robert Wiebe, *The Search for Order, 1877–1920* (New York: Hill and Wang, 1967); T. J. Jackson Lears, *No Place of Grace: Antimodernism and the Transformation of American Culture, 1880–1920* (New York: Pantheon, 1981).

8. Dr. Charles Chandler, a chemist working for New York City's Health Department, began laboratory analysis of milk in 1870, but he was only able to determine whether the milk had been adulterated with excess water. By 1900, Dr. William H. Park conducted laboratory examinations of milk in New York City that determined not only if the milk had been watered, but also whether it carried pathogenic bacteria. Duffy, *The Sanitarians*, 185; Charles E. North, "Milk and Its Relation to Public Health," in *A Half-Century of Public Health*, ed. Mazcyck Ravenel, 237–289 (New York: American Public Health Association, 1921); Richard Meckel, *Save the Babies: American Public Health Reform and the Prevention of Infant Mortality* (Baltimore: Johns Hopkins University Press, 1990), 39.

9. Because of the close relationship between agricultural scientists, the farm press, and business leaders, throughout this chapter I use the term "agricultural reformers" to refer broadly to those who promoted new agricultural techniques. For the history of agricultural science, see Roy V. Scott, *The Reluctant Farmer: The Rise of Agricultural Extension to 1914* (Urbana: University of Illinois Press, 1970); Alan Marcus, *Agricultural Science and the Quest for Legitimacy* (Ames: Iowa State University Press, 1985); Charles Rosenberg, *No Other Gods: On Science and American Social Thought*, rev. ed. (Baltimore: Johns Hopkins University Press, 1997), 153–199; Deborah Fitzgerald, *The Business of Breeding: Hybrid Corn in Illinois, 1890–1940* (Ithaca: Cornell University Press, 1990), 1–2; and Fitzgerald, *Every Farm a Factory: The Industrial Ideal in American Agriculture* (New Haven: Yale University Press, 2003).

10. Samuel P. Hays, *Conservation and the Gospel of Efficiency* (Cambridge: Harvard University Press, 1959).

11. The classic work on wilderness preservation is Roderick Nash, *Wilderness and the American Mind*, 3rd ed. (New Haven: Yale University Press, 1982); see also William Cronon, "The Trouble with Wilderness; or, Getting Back to the Wrong Nature," in *Uncommon Ground: Rethinking the Human Place in Nature*, ed. William Cronon, 69–90 (New York: Norton, 1995).

12. By 1894, 88.9 percent of Chicago's milk supply came from outside of the city. Daniel Ralston Block, 'Development of Regional Institutions in Agriculture: The Chicago Milk Marketing Order' (Ph.D. diss., University of California, Los Angeles, 1997), 46. Boston's milk came from as far as 243 miles away, New York City's from 400 miles away. Milton J. Rosenau, *The Milk Question* (Boston: Houghton Mifflin, 1912), 260; J. Scott Macnutt, *The Modern Milk Problem* (New York: Macmillan, 1917), 36–44. Still, many rural families obtained milk from a family cow. Susan Matthews, "Food Habits of Georgia Rural People," Bulletin No. 159 (Athens: Georgia Experiment Station, 1929), 26; J. O. Rankin, "Cost of Feeding the Nebraska Farm Family," Bulletin 219 (Lincoln: University of Nebraska Agricultural Experiment Station, 1927), 8, 15, 17.

13. Rosenau, *The Milk Question,* 20.

14. Wolf, *Don't Kill Your Baby*, 10–11; Meckel, *Save the Babies*, 49–50.

15. J. H. Mason Knox Jr., "Speech Delivered to Johns Hopkins School of Nursing Alumnae, Baltimore, Maryland, 1906," reprinted as "The Relation of the Milk Supply to Infant Mortality," *American Journal of Public Health,* 89 (March 1999): 408.

16. Apple, *Mothers and Medicine*, 16–18; Meckel, *Save the Babies*, 57.

17. John Spargo, *The Common Sense of the Milk Question* (New York: Macmillan, 1908), 31.

18. Wolf, *Don't Kill Your Baby*, 9–41.

19. Comments from a Conference on Prevention of Infant Mortality, Yale University, November 11 and 12, 1909, printed in "Society Transactions," *Pediatrics* 22 (January 1910): 74, 76.

20. Wolf, *Don't Kill Your Baby*, 76–77, 108–109, 111–126; Meckel, *Save the Babies*, 96.

21. Wolf, *Don't Kill Your Baby*, 21–22.

22. Levenstein, "'Best for Babies' or 'Preventable Infanticide'?" 80–81; Meckel, *Save the Babies*, 55, 58.

23. Agricultural census takers found only 5.6 percent of dairy cows in the United States in 1900 and 1910 lived off of farms. *Thirteenth Census of the United States Taken in the Year 1910. Vol. 5: Agriculture* (Washington, D.C.: Government Printing Office, 1913), 455, table 73; Meckel, *Save the Babies*, 63.

24. Wolf, *Don't Kill Your Baby*, 51–52; Dupuis, *Nature's Perfect Food*, 18–19.

25. W. A. Henry, "Comparative Value of Feeds, with Tables Giving Their Percentage of Digestible Nutrients," in *Creamery Patron's Handbook* (Chicago: National Dairy Union, 1902), 39.

26. *Fischer v. St. Louis*, 194 U.S. 361; 24 S. Ct. 673; 48 L.Ed. 1018; 1904 U.S. LEXIS 830, argued April 12, 1904; see also Rosenau, *The Milk Question*, 260–261.

27. Paul Boyer, *Urban Masses and Moral Order in America, 1820–1920* (Cambridge: Harvard University Press, 1978), 92–93, 279–280.

28. Nancy Tomes, *Gospel of Germs: Men, Women, and the Microbe in American Life* (Cambridge: Harvard University Press, 1998), 152–153.

29. For working-class milk consumption, see Louise Bolard More, *Wage-Earners' Budgets: A Study of Standards and Cost of Living in New York City* (New York: Henry Holt, 1907), 209; R. C. Chapin, *The Standard of Living among Workingmen's Families in New York City* (New York: Charities Publication Committee, 1909), 135; on branding, Susan Strasser, *Satisfaction Guaranteed: The Making of the American Mass Market* (Washington, D.C.: Smithsonian Books, 1989), 52–53, 204–205; Hal Barron, *Mixed Harvest: The Second Great Transformation in the Rural North, 1870–1930* (Chapel Hill: University of North Carolina Press, 1997), 81–105.

30. Century Milk Company Advertisement, *New York Times*, April 22, 1901, April 24, 1901.

31. Mary Mercer, "What Housewives' Leagues Are Doing: New England Conference of House-wives," *Housewives' League Magazine* 9 (May 1917): 20–21; Mrs. C. D. Allin, *Housewives' League Magazine* 8 (November 1916): 41; "Must We Pay More for Bread and Milk?" *Housewives' League Magazine* (October 1916): 42.

32. For examples of the images, see "Moo Cow Moo," "I'll Wait for You Till the Cows Come Home," "Mollly Drive the Cows Home," "Driving the Cows From Pasture," "On the Meadow," and "The Country Girl," Samuel de Vincent Sheet Music Collection, Accession 300, Box 488A, Folder Cows, Archives Center, National Museum of American History, Smithsonian Institution, Washington, D.C. (hereafter de Vincent Collection, Archives Center, NMAH); Blue Valley Creamery Company postcard, Warshaw Collection of Business Americana, Accession 60, Dairy, Box 1, unnumbered folder. Archives Center, NMAH; Vermont Farm Machine Company Trade Catalog, *United States Cream Separators* (Bellows Falls, Vt.: Vermont Machine Company, 1910), 1–3.

33. Jackson Lears, *Fables of Abundance: A Cultural History of Advertising in America* (New York: Basic, 1994), 18, 110, 122.

34. David Stradling, *Making Mountains: New York City and the Catskills* (Seattle: University of Washington Press, 2007), 25, 107.

35. Julie Miller, "To Stop the Slaughter of the Babies: Nathan Straus and the Drive for Pasteur-ized Milk, 1893–1920," *New York History* (April 1993): 172, 177–179.

36. Gregg Mitman, "In Search of Health: Landscape and Disease in American Environmental History," *Environmental History* 10 (2005): 197; Sheila Rothman, *Living in the Shadow of Death: Tuberculosis and Social Experience of Illness in American History* (Baltimore: Johns Hopkins University Press, 1994), 45–56, 131–147; Michael Teller, *The Tuberculosis Movement: A Public Health Campaign in the Progressive Era* (New York: Greenwood, 1988); Susan Craddock, *City of Plagues: Disease, Poverty, and Deviance in San Francisco* (Minne-apolis: University of Minnesota Press, 2000), 216–218.

37. Henry Alvord, "Utilization of By-Products of the Dairy," in U.S. Department of Agriculture, *Annual Yearbook of Agriculture* (Washington, D.C.: Government Printing Office, 1898), 528; Junket Cookbook, Product Cookbooks Collection, Archives Center, National Museum of American History, Smithsonian Institute, Washington, D.C.; Fannie Merritt Farmer, *Food and Cookery for the Sick and Convalescent* (Boston: Little, Brown, 1904), 50; "Feeding America: The Historic American Cookbook Project," Michigan State University, Special Collections, available at http://digital.lib.msu.edu/projects/cookbooks/html/books/book_56.cfm.

38. B. Meade Browne, "Sanitary Water Supplies for Dairy Farms," in *Milk and Its Relation to Public Health,* Hygienic Laboratory Bulletin 56, March 1909 (Washington, D.C.: Government Printing Office, 1912); Parker, *City Milk Supply,* 131, 167; Meckel, *Save the Babies,* 77. Dairies that fell short of the conditions of the dairy as idealized were more diffi-cult to hide in places like Los Angeles, where most of the city's milk came from farms located close to the central city. Folder I/2/20, Box 1, California Dairy Industry History Collection, California State Parks Archive, Sacramento, Calif.

39. Tomes, *Gospel of Germs,* 52–53, 57; Linda Nash, *Inescapable Ecologies: A History of Environ-ment, Disease, and Knowledge* (Berkeley: University of California Press, 2007), 89–91.

40. Spargo, *Common Sense of the Milk Question,* 313.

41. Parker, *City Milk Supply,* 373, 382–385; Archibald Robinson Ward, *Pure Milk and the Public Health: A Manual of Milk and Dairy Inspection* (Ithaca, N.Y.: Taylor and Carpenter, 1909), 195–196; Wm. Creighton Woodward, "The Municipal Regulation of the Milk Supply of the District of Columbia," in *Milk and Its Relation to Public Health,* 793–809, 818–821; Anonymous, "Can Revoke Milk Licenses," *New York Times,* May 6, 1906, p. 9; James Tobey, ed., *Public Health Law: A Manual of Law for Sanitarians* (Baltimore: Williams and Wilkins, 1926), 99.

42. James D. Brew, *Milk Quality as Determined by Present Dairy Score Cards,* Bulletin 398 (Ithaca, N.Y.: New York Agricultural Experiment Station, 1915), 111; W. K. Brainerd and

W. L. Mallory, "Milk Standards: A Study of the Bacterial Count and Dairy Score Card in City Milk Inspection," Virginia Polytechnical Institute Agricultural Experiment Station, Bulletin 194 (September 1911), 18; MacNutt, *The Modern Milk Problem*, 70–71; Rosenau, *The Milk Question*, 171; Spargo, *Common Sense of the Milk Question*, 313–318.

43. Parker, *City Milk Supply*, 116; Frank White and Clyde I. Griffith, *Barns for Wisconsin Dairy Farms*, Bulletin 266 (Madison: University of Wisconsin Agricultural Experiment Station Bulletin, 1916), 9–13.

44. Sarah D. Belcher, *Clean Milk* (New York: Hardy, 1903), 33–36.

45. Belcher, *Clean Milk*, 31–32; Barbara Gutman Rosenkrantz, *Public Health and the State: Changing Views in Massachusetts, 1842–1936* (Cambridge: Harvard University Press, 1972), 108; agricultural reformers exhorted farmers to apply manure regularly to farm fields, not just to prevent it from spreading germs or contaminating milk, but because the longer manure sat in the barnyard, the less its potential fertilizing value.

46. Rosenau, *The Milk Question*, 77; Peter Atkins, *Liquid Materialities: A History of Milk, Science, and the Law* (Burlington, Vt.: Ashgate, 2010), 233–234.

47. North, in MacNutt, *The Modern Milk Problem*, 161.

48. Woodward, "Municipal Regulation," 774, 776; Brew, "Milk Quality," 110; MacNutt, *The Modern Milk Problem*, 85; Parker, *City Milk Supply*, 177.

49. Brainerd and Mallory, "Milk Standards," 7, 20; Rosenau, *The Milk Question*, 76; C. V. Craster, "The Value of Municipal Dairy Inspection," *American City* 23 (1920): 172–175.

50. Folder I/2/20, Dairy Roadside Appearance Program files, box 1, California Dairy Industry History Collection, California State Parks Archive, Sacramento, Calif.

51. Parker, *City Milk Supply*, 455; Brew, "Milk Quality," 130.

52. For first interpretation, see Dupuis, *Nature's Perfect Food*, 73, 84–89, 130; David Danbom, *The Resisted Revolution: Urban America and the Industrialization of Agriculture, 1900–1930* (Ames: Iowa State University Press, 1979), 47, 144; Barron, *Mixed Harvest*, 87–90. For the latter, see Thomas Pegram, "Public Health and Progressive Dairying in Illinois," *Agricultural History* 65 (1991): 36–50; Barbara Gutman Rosenkrantz, "The Trouble with Bovine Tuberculosis," *Bulletin of the History of Medicine* 59 (Summer 1985): 155–175; Leavitt, *The Healthiest City*, 156–189; Paula Baker, *Moral Frameworks of Public Life: Gender, Politics, and the State in Rural New York, 1870–1930* (New York: Oxford University Press, 1991), 131–133.

53. George Whitaker, *The Milk Supply of Chicago and Washington*, Bulletin 438 (Washington, D.C.: U.S. Department of Agriculture, Bureau of Animal Industry, 1911), 33.

54. Pennsylvania State Archives, Harrisburg; Record Group 1, Records of the Department of Agriculture; Division of Crop Reporting; Farm Census Returns, 1927, *Bradford County*.

55. O. R. Johnson and R. M. Green, *Profits from Milk Cows on General Cornbelt Farms*, Bulletin 159 (Columbia: University of Missouri Press, 1919), 7.

56. Manfred Waserman, "Henry L. Coit and the Certified Milk Movement in the Development of Modern Pediatrics," *Bulletin of the History of Medicine* 46 (1972): 364–390; Meckel, *Save the Babies*, 102–104; Levenstein, "Best for Babies or 'Preventable Infanticide?'" 85–86; Rosenau, *The Milk Question*, 141–169; MacNutt, *The Modern Milk Problem*, 67–69.

57. MacNutt, *The Modern Milk Problem*, 147, 151.

58. H. A. Harding, "Publicity and Payment Based on Quality as Factors in Improving a City Milk Supply," Bulletin 337 (Geneva: New York State Agricultural Experiment Station, 1911), 104, 114; H. A. Harding and James Brew, "Financial Stimulus in City Milk Production," Bulletin 363 (Geneva: New York State Agricultural Experiment Station, 1913), 170, 174.

59. Mary Mercer, "What Housewives' Leagues Are Doing—New England Conference of Housewives," *Housewives' League News* 9 (May 1917): 20–21; Glickman, *Buying Power*, 157, 174–184.

60. Description of farm equipment in correspondence from Franklin Pope Wilson to Mutual Fire Insurance Company, January 15, 1917; Box 1, Folder Business Correspondence, 1917, Franklin Pope Wilson Papers, Mss 1W6933a FA 2, Virginia Historical Society, Richmond, Va. (hereafter, Franklin Pope Wilson Papers, VHS).

61. Franklin Wilson to Charles Arnett, January 1, 1922, February 19, 1922, Box 1, Folder Business Correspondence Oak Knoll Farm, 1922; Franklin Pope Wilson Papers, VHS.

62. Charles Arnett to Franklin Wilson, April 4, 1928, Folder Business Correspondence, Oak Knoll Farm, 1927–1928; Franklin Wilson to Thompson's Dairy, June 8, 1923, Folder Business Correspondence, Thompson's Dairy 1916–1926; both in Franklin Pope Wilson Papers, VHS.

63. Harding and Brew, "Financial Stimulus in City Milk Production," 175–177.

64. *Littlefield v. State of Nebraska*, 42 Neb. 223, 60 N.W. 724 (1894).

65. George Whitaker, "The Extra Cost of Producing Clean Milk," in U.S. Department of Agriculture, *Twenty-Sixth Annual Report of the Bureau of Dairy Industry, 1909* (Washington, D.C.: Government Printing Office, 1911), 121–130.

66. W. Nicholas Lackey, "Pediatric Practice in the Small Towns and Country," *Pediatrics* 25 (June 1913): 370, 374; K. E. Miller, "Safe Milk for the Small Town," *Public Health Reports* (December 1918): 2213–2215; Parker, *City Milk Supply*, 476–480.

67. Rosenberg, *No Other Gods*, 186.

68. The U.S. population increased by 20 percent between 1900 and 1910, but the increase in urban population outpaced rural areas. Between 1910 and 1920, the total U.S. population increased 14.9 percent, but while the rural population increased 3.2 percent, the urban population increased by 28.8 percent. U.S. Bureau of the Census, *Thirteenth Census of the United States, vol. 5: Agriculture,* 27; *Fourteenth Census of the United States, Taken in the Year 1920, Volume VI, Part 2—Agriculture* (Washington, D.C.: Government Printing Office, 1922) 17; William Bowers, *The Country Life Movement in America, 1900–1920* (Port Washington, N.Y.: Kennikat Press, 1974), 3–6; for examples of country life ideas in the dairying press, see "Farm School Bill Signed," *Pacific Dairy Review* 11, no. 5 (February 28, 1907): 1.

69. Rosenkrantz, "The Trouble with Bovine Tuberculosis," 167. Recent estimates claim 10 percent of dairy animals were infected in 1917. Alan Olmstead and Paul Rhode, "Not on My Farm! Resistance to Bovine Tuberculosis Eradication in the United States," *Journal of Economic History* 67 (September 2007): 772.

70. Veranus Alva Moore, *Bovine Tuberculosis and Its Control* (Ithaca, N.Y.: Carpenter and Company, 1913), 22–23, 29–32; E. C. Schroeder, "The Relation of the Tuberculous Cow to Public Health," in *Milk and Its Relation to Public Health*, 533.

71. Teller, *The Tuberculosis Movement*, 67; Alan Olmstead and Paul Rhode, "The 'Tuberculous Cattle Trust': Disease Contagion in an Era of Regulatory Uncertainty," *Journal of Economic History* 64 (December 2004): 934.

72. Mazyck Ravenel, "Relations of Human and Bovine Tuberculosis," Sixth International Congress on Tuberculosis, *Transactions*, vol. 4, pt. 2, sec. 7, *Tuberculosis in Animals and Its Relation to Man* (Philadelphia: Fell Company, 1908), 683; Rosenkrantz, "The Trouble with Bovine Tuberculosis," 156–157; Olmstead and Rhode, " 'The 'Tuberculous Cattle Trust,' " 934–935.

73. Henry Shumway, *A Hand-book on Tuberculosis among Cattle with Considerations of the Relation of the Disease to the Health and Life of the Human Family, and of the Facts Concerning the Use of the Tuberculin as a Diagnostic Test* (London: Sampson Low, Marston, and Company, 1895), 25; H. L. Russell, "Bovine Tuberculosis and the Tuberculin Test," *University of Wisconsin Agricultural Experiment Station*, Bulletin 40 (July 1894), 9–18; H. B. Leonard, "The Subcutaneous Test," in *Proceedings of the Tuberculosis-Eradication Conference of State and Federal Livestock Sanitary Officials* (Chicago, October 6–8, 1919), 18–19.

74. Wm. Leonard in testimony from the *Adams v. Milwaukee* trial, in Joint Committee on the Tuberculin Test and Pasteurization of Milk, *Evidence Taken before the Joint Committee on the Tuberculin Test, 1911: Volume I* (Springfield: Illinois State Journal Company, 1912), 542.

75. Leonard Pearson and Mazyck Ravenel, *Tuberculosis of Cattle and the Pennsylvania Plan for Its Repression*, Bulletin 75, Commonwealth of Pennsylvania Department of Agriculture (Harrisburg, Pa.: William Stanley Ray, 1901), 326.

76. T. A. Green to H. R. Russell, December 11, 1909; University of Wisconsin College of Agriculture, Administration: Office of the Dean and Director; General Files, Russell Papers; H. L. Russell, Miscellaneous Files, Series 9/1/1/22–4, box 3, University of Wisconsin Archives, Steenbock Library, Madison, Wisc., Box 3, UW-Archives (hereafter referred to as Russell Papers).

77. Leonard, in *Evidence Taken before the Joint Committee on Tuberculin Test*, 542.

78. U.S. Department of Agriculture, Bureau of Animal Industry, *Special Report on Diseases of Cattle* (Washington, D.C.: Government Printing Office, 1904), 405.

79. H. L. Russell, "The Function of the State in Milk Control," *Creamery and Milk Plant Monthly* 1 (October 1912): 8; Wisconsin Dairymen's Association, *Thirty-Sixth Annual Report of the Wisconsin Dairymen's Association* (Madison: Democrat Printing Company, 1910), 191.

80. Waserman, "Henry L. Coit and the Certified Milk Movement," 380; John Gould, "Note and Comment," *Hoard's Dairyman* 31 (February 16, 1900): 12–13.

81. Farm families who regularly obtained whey or skim milk from dairy plants for feeding purposes could also introduce the disease to their animals through the pooled skim milk. Moore, *Bovine Tuberculosis and Its Control*, 49; Harry Russell, *The Spread of Tuberculosis through Factory Skim Milk with Suggestions as to Its Control*, Bulletin 143 (Madison: University of Wisconsin Experiment Station, 1907), entire; A. J. Glover, ed., *Thirty-Sixth Annual Report of the Wisconsin Dairymen's Association* (Madison: Democrat Printing Company, 1908), 195–196.

82. Leavitt, *The Healthiest City*, 183–187; Frank DiLeva, "Frantic Farmers Fight Law," *Annals of Iowa* 32 (October 1953): 81–109; Dupuis, *Nature's Perfect Food*, 73; Susan Jones, *Valuing Animals: Veterinarians and Their Patients in Modern America* (Baltimore: Johns Hopkins University Press, 2003), 70.

83. Shirley Schlanger Abrahamson, "Law and the Wisconsin Dairy Industry: Quality Control of Dairy Products, 1838–1929" (Ph.D. diss., University of Wisconsin-Madison, 1962), 6–13; Olmstead and Rhode, "Not on My Farm!" 776; Harding and Brew, "Financial Stimulus in Milk Production," 179.

84. The most celebrated challenge to tuberculin testing reached the U.S. Supreme Court. *Adams v. City of Milwaukee*, 228 U.S. 578. Other cases included *St. Louis v. Liessing* (1905) 190 Mo. 464; *New Orleans v. Caroleau* (1908) 121 La. 890; *Nelson v. Minneapolis* (1910) 112 Minn. 16; *Borden v. Montclair* (1911) 81 NJL 218; *Hawkins v. Hoye* (1914) 108 Miss. 282. For opposition to tuberculosis testing, see C. W. Jennings, "Penciled Pot-Pourri," *Hoard's Dairyman* 31 (May 25, 1900): 308; Charles Chapin, *Municipal Sanitation in the United States* (Providence: Snow and Farnham, 1901), 771; Dileva, "Frantic Farmers Fight Law," 90–103; Olmstead and Rhode, "Not on My Farm!," 768–809.

85. Wolf, *Don't Kill Your Baby*, 59–61; Pegram, "Public Health and Progressive Dairying in Illinois," 36–50.

86. Pearson and Ravenel, *Tuberculosis of Cattle and the Pennsylvania Plan for Its Repression*, 198.

87. Jos. Kitchener to H. L. Russell, April 2, 1909, Russell Papers; A. J. Glover, *Thirty-Eighth Annual Report of the Wisconsin Dairymen's Association* (Madison: Democrat Publishing Company, 1910), 157–159; H. W. Griswold to H. L. Russell, January 11, 1909, Box 3, Russell Papers; see, for instance, the influence of the Bureau of Animal Industry in U.S. Treasury Department, *Milk and Its Relation to the Public Health*, 499–553; Sixth International Congress on Tuberculosis, *Transactions*, vol. 4, pt. 2, sec. 7; Jones, *Valuing Animals*, 77–85.

88. H. R. Russell to P. V. Collins Publishing Company, January 11, 1911; Box 3, Russell Papers.

89. Elihu Gifford Papers, Box 1, Diary 1923–1931, Elihu Gifford Papers, SC 18734, New York State Manuscripts and Special Collections, Albany, N.Y.

90. Miller, "To Stop the Slaughter of the Babies," 163.

91. Meckel, *Save the Babies*, 87; Anne Baber Kennedy, "J. H. Mason Knox," *American Journal of Public Health* 89 (March 1999): 409–410.

92. W. H. Park and L. Emmett Holt, "Pure and Impure Milk in Infant Feeding," *Pediatrics* 15 (December 1903): 724, 732–733.

93. H. A. Harding, "Compulsory Tuberculin Tested Milk Versus Compulsory Pasteurization for City Milk Supplies," *Creamery and Milk Plant Monthly* 3 no. 3 (November 1914), 21–22. Leslie Lumsden, "The Milk Supply of Cities in Relation to the Epidemiology of Typhoid Fever," in *Milk and Its Relation to Public Health*, 164.

94. Meckel, *Save the Babies*, 82.

95. Alfred Hess, "Infantile Scurvy," *Pediatrics* 7, no. 7 (July 1916): 305, 308–309; Spargo, *Common Sense of the Milk Question*, 255–259.

96. Parker, *City Milk Supply*, 269.

97. Wasserman, "Henry L. Coit and the Certified Milk Movement," 373.

98. Editorial, "Pasteurized or Raw?" *Pacific Dairy Review* 11 (May 9, 1907): 2.

99. Marion Campbell, "The Cream Top Bottle," 2–3, box 488A, de Vincent Collection, Archives Center, NMAH.

100. Davis Milk Machinery Company, Advertisement, *Creamery and Milk Plant Monthly* 5, no. 1 (September 1916): 7.

101. Charles Kilbourne, "What Causes Contribute to the Loss of the Cream Line in Bottles," *Creamery and Milk Plant Monthly* 3 (April 1915): 11–20.

102. Parker, *City Milk Supply*, 312–313; Olmstead and Rhode, "An Impossible Undertaking," 736.

103. Dudley Harmon, "The Truth about Milk," *Ladies Home Journal* 35, no. 3 (March 1918): 44; Philip Hawk, "What We Eat: And What Happens to It, The Truth about the Milk We Drink!" *Ladies Home Journal* 4, no. 2 (February 1917): 29; National Dairy Council, "Milk Is a Food, Not a Beverage," *Creamery and Milk Plant Monthly* 5, no. 11 (July 1917): 16; S. Josephine Baker, "The 'In-Between' Child in Wartime," *Ladies Home Journal* 35, no. 6 (June 1918): 1.

104. Harvey Wiley, "Milk," *Good Housekeeping* 35, no. 2 (February 1918): 49.

105. The phrase was used as early as 1907 but was popularized in the 1920s. Louden Machinery Company, *Some Interesting Facts on a Homely Subject* (Fairfield, Iowa: Louden Machinery Company, 1907), 15; Hawk, "What We Eat," 29. By the 1920s, the Eskimo Pie company painted the phrase "nature's perfect food" on its delivery trucks. Eskimo Pie Collection, 1921–1992, Series 2: Historical and Background Information, Subseries 1: Background information on the company, box 6, folder 2: articles re: Eskimo Pie, Archives Center, NMAH, Smithsonian Institution. See also Samuel Crumbine and James Tobey, *The Most Nearly Perfect Food: The Story of Milk* (Baltimore: Williams and Wilkins, 1929).

106. George M. Whitaker, "Food Materials in Milk," *Hoard's Dairyman* 41 (September 2, 1905): 905; "Where the Silo Is Popular," *Pacific Dairy Review* 11, no. 21 (June 20, 1907): 1.

Chapter 2

1. Mrs. Julian Heath, "What Happens When Women Band Together to Demand Reform," *New York Times*, November 10, 1912.

2. Economic Research Service, USDA, "Food Availability Data," http://www.ers.usda.gov/ Data/FoodConsumption (accessed March 24, 2009). In 1924, C. W. Larson, the chief of the Division of Dairying in the USDA, reported that working families consumed 89.7 pounds of butter per family each year. Ethelbert Stewart, "Economics of Creamery Butter Consumption," *Monthly Labor Review* 21 (July 1925): 1–2. In some locales, butter consumption was even higher. In 1927, for instance, an average Milwaukee family consumed 122.2 pounds of packaged butter per year. Marvin Schaars, "The Butter Industry of Wisconsin" (Ph.D. diss., University of Wisconsin-Madison, 1932), 203.

3. Response to A.A., *Housewives' League News* 11 (January 1918): 32.

4. Irene Till, *Milk: The Politics of an Industry* (New York: McGraw-Hill, 1938), 462.

5. Steven Keillor, *Cooperative Commonwealth: Co-ops in Rural Minnesota, 1859–1939* (St. Paul: Minnesota State Historical Society Press, 2000), 293–296.

6. William Leuchtenburg, *The Perils of Prosperity, 1914–1932*, 2nd ed. (Chicago: University of Chicago Press, 1993), 6–7; Glen Gendzel, "1914–1929," in *A Companion to 20th-Century America*, ed. Stephen Whitfield, 29–30 (Malden, Mass.: Blackwell, 2004).

7. Kenneth Ruble, *Men to Remember: How 100,000 Neighbors Made History* (Chicago: R. R. Donnelley and Sons, 1947), 29.

8. Tom B. Cunningham, interview by Lu-Ann Jones, January 9, 1987, Southern Agriculture Oral History Project Records, box 16, folder 1, Archives Center, National Museum of American History, Smithsonian Institution, Washington, D.C. (hereafter Archives Center, NMAH); Lu Ann Jones, *Mama Learned Us to Work: Farm Women in the New South* (Chapel Hill: University of North Carolina Press, 2002), 12, 50, 78.

9. Whereas today's dairy manufacturing plants often incorporate cheese, butter, ice cream, yogurt, and milk bottling under one roof, in the 1910s and 1920s, these processes usually took place in separate factories: creameries, cheese factories, milk condensaries, and milk bottlers. Some creameries also manufactured ice cream or bottled milk, but the use of the term "creamery" is specific to the butter trade.

10. Theodore Macklin, "A History of the Organization of Creameries and Cheese Factories in the United States" (Ph.D. diss., University of Wisconsin-Madison, 1917), 15–20; Loyal Durand Jr., "The Lower Peninsula of Michigan and the Western Michigan Dairy Region: A Segment of the American Dairy Region," *Economic Geography* 27 (1951): 163–164. As late as 1913, 87 percent of butter in northern Minnesota was made at home. Louis D. H. Weld, *Social and Economic Survey of a Community in the Red River Valley* (Minneapolis: University of Minnesota Press, 1915), 27.

11. James Mayo, *The American Grocery Store: The Business Evolution of an Architectural Space* (Westport, Conn.: Greenwood, 1993), 82, 84; Tracey Deutsch, *Building a Housewife's Paradise: Gender, Politics, and American Grocery Stores in the Twentieth Century* (Chapel Hill: University of North Carolina Press, 2010), 58–59.

12. Elihu Gifford Papers, Diaries 1911–1931, box 1, items (hereafter Gifford papers, NYSL); *History of Oneida County New York: From 1700 to the Present Time*, vol. 2 (Chicago: S. J. Clarke Publishing, 1912), 397–398.

13. For feed rations, see *Livestock Feed Journal*, 1913–1942, box 3, item 1 and Diary, August 1916–August 1923, box 1, item 6; for cream sales and lemons and pineapple purchases, see *Journal*, November 1918–July 1931, box 2, item 4, all in Elihu Gifford Papers, SC 18734, New York State Manuscripts and Special Collections, Albany, N.Y.

14. G. L. McKay and C. Larson, *Principles and Practice of Buttermaking*, 2nd ed. (New York: Wiley, 1908), 150–160; Edward Weist, *The Butter Industry in the United States: An Economic Study of Butter and Oleomargarine* (New York: Columbia, 1916), 23.

15. McKay and Larsen, *Principles and Practice of Buttermaking*, 155.

16. Sharples Cream Separator Advertisement, n.d., Warshaw Collection of Business Americana, Dairy, box 1, folder 24, Archives Center, NMAH; a single issue of *Hoard's Dairyman* carried ads from cream separator companies De Laval, Vermont Farm Machinery Company, Galloway, Sharples, Albaugh-Dover, American Separator, Smith Manufacturing, and International Harvester, *Hoard's Dairyman* 41 (February 4, 1910): 2, 4–6, 17.

17. Gifford Diaries, November 3, 1910, item 4, SC 18734, NYSL.

18. Otto Hunziker, *The Butter Industry*, 3rd ed. (LaGrange, Ill.: Hunziker, 1940), 14.

19. Macklin, "A History of the Organization of Creameries and Cheese Factories in the United States," 115, 122; C. H. Benkendorf, "The Separator Problem: Difficulties between Creameries and Patrons and How to Handle Them," *Pacific Dairy Review* 11, no. 9 (March 28, 1907): 10–13; Newton Clark, "Butter Prices from Producer to Consumer," *Bulletin of the Bureau of Labor Statistics* (Washington, D.C.: Government Printing Office, 1915): 16–19; H. A. Harding, "Relation of Bacteria in Dairy Products to Health," *Creamery and Milk Plant*

Monthly 5, no. 3 (November 1916): 38–43; Hunziker, *The Butter Industry*, 16–17; Aaron Ihde, interview by William K. Laderfer, May 15, 1963, interview transcript 263, University of Wisconsin-Madison Archives, Madison, Wisc.; Floyd Lucia interview, by Dale Treleven, January 16, 1976, tape 1, side 1, interview M75–34. Wisconsin Agriculturalists Oral History Project, Wisconsin Historical Society, Madison, Wisc. (hereafter WAOHP, WHS).

20. Weld, *Social and Economic Survey*, 46; Levi Wells, "Renovated Butter: Its Origin and History," U.S. Department of Agriculture, *Annual Yearbook of Agriculture* (Washington, D.C.: Government Printing Office, 1905), 395–396; Hunziker, *The Butter Industry*, 581–582.

21. Daniel Rodgers, *Atlantic Crossings: Social Politics in a Progressive Age* (Cambridge: Harvard University Press, 1998), 321–327; Hal Barron, *Mixed Harvest: The Second Great Transformation in the Rural North, 1870–1930* (Chapel Hill: University of North Carolina, 1997), 107–108.

22. Isabel Baumann, interview by Dale Treleven, April 1, 1980, tape 1, side 1, WAOHP, WHS; Clara Weiskirsher Scott, interview by Jean Sauls Rannels, February 7, 1985, Rural Women's Oral History Project, box 1, folder 4, Mss 719, WHS; Elizabeth Sullivan, interview by Dale E. Treleven, May 11, 1977, WAOHP, WHS; Ruth Schwartz Cowan, *More Work for Mother: The Ironies of Household Technology from the Open Hearth to the Microwave* (New York: Basic, 1983).

23. Rangar and Margaret Segerstrom, interview by Dale Treleven, September 29, 1976, WAOHP, WHS; Jens Christensen, Withee, Wisc., December 11, 1913; Grant Burlingame, Ripon, Wisc., January 13, 1914; Herman Buchholtz, Weyanmego, Wisc., February 12, 1914. Wisconsin Dairy and Food Division, Milk Adulteration Prosecution Reports, 1909–1950, box 1, WHS.

24. Gifford Papers, box 1, item 4, SC 18734, NYSL W. S. Greene, "The Production and Examination of Cream for Buttermaking," 2, box 2204, f. Milk. Records of the Office of the Secretary of Agriculture, General Correspondence, 1906–1970, RG 16, SOA, GC, 1906–70, NARA.

25. Greene, "The Production and Examination of Cream for Buttermaking," 6, 8, RG 16, SOA, GC, 1906–70, NARA.

26. Aaron Ihde, interview, University of Wisconsin-Archives. See also Otto Hunziker, "Answer to Question Submitted by Urner-Barry Company, New York," 2, September 28, 1924. Otto Hunziker Papers, M68–235, Wisconsin Historical Society Archives, Madison, Wisc. (Hunziker was the chief chemist at Blue Valley Creamery and Ihde's supervisor.)

27. Floyd B. Lucia, interview tape 1, side 1, WHS.

28. Greene, "The Production and Examination of Cream for Buttermaking," 3; Hunziker, "Answer to Question Submitted by Urner-Barry Company, New York."

29. Hunziker, *The Butter Industry*, 15; E. S. Larrabee and G. Wilster, "The Butter Industry of Oregon: A Study of Factors Relating to the Quality of Butter," *Oregon State Agricultural College Bulletin 258* (Corvallis: Oregon State Agricultural College, December 1929): 20.

30. Lucia, tape 1, side 2.

31. McKay and Larsen, *Principles and Practice of Buttermaking*, 93.

32. M. P. Mortenson to A. J. McGuire, February 23, 1926, box 2, folder 6, A. J. and Marie McGuire Papers, P929, Minnesota Historical Society, St. Paul, Minnesota, (hereafter McGuire Papers, MNHS).

33. McGuire Papers, box 1, MNHS.

34. McGuire Papers, box 2, MNHS.

35. Oakdale Cooperative Butter Association—Minutes from January 15, 1916, meeting, box 6, folder 10, WI Dairies Coop, Mss 473, WHS.

36. A. J. McGuire, "The Farmer and the Middleman" (1915–1920), A. J. McGuire Papers, box 1., MNHS.

37. H. Colin Campbell, "Concrete Roads and Milk," *Hoard's Dairyman* 62 (October 21, 1921): 397; J. H. Brown, "Automobiles on Dairy Farms," *Hoard's Dairyman* 41 (September 9, 1910): 945.

38. Schaars, "The Butter Industry of Wisconsin," 73.

39. Hunziker, *The Butter Industry*, 351.

40. McKay and Larsen, *Principles and Practice of Buttermaking*, 184, 192–199; Otto Hunziker, "Neutralization of Cream for Buttermaking," paper given at the Annual Meeting of American Association of Creamery Butter Manufacturers, February 19, 1918, Chicago, Ill. Hunziker Papers, box 1, Accession M68–235, WHS.

41. "Neutralizer and Adulterated Butter," *Hoard's Dairyman* 62 (February 25, 1921): 226.

42. William White, "Dairy Earnings Larger If Cream Is Marketed While Fresh and Sweet," in U.S. Department of Agriculture, *Annual Yearbook of Agriculture* (Washington, D.C.: Government Printing Office, 1931), 187–188.

43. William White, "Dairy Market Finds Sweet-Cream Butter Gains Public Favor," in *United States Department of Agriculture, Yearbook of Agriculture, 1927* (Washington, D.C.: Government Printing Office, 1927), 266; Edward Sewall Guthrie, *The Book of Butter: A Text on the Nature, Manufacture, and Marketing of the Product* (New York: Macmillan, 1918), 109.

44. Weist, *The Butter Industry*, 151–152; Guthrie, *The Book of Butter*, 186–187; Ruble, *Men to Remember*, 40–41.

45. Hunziker, "Neutralization of Cream for Buttermaking," 11; G. L. McKay, "Facts about Butter: Suggestions for a Standard," *Bulletin 12* (Chicago: American Creamery Butter Manufacturers, June 1918), 22–23.

46. Guthrie, *The Book of Butter*, 88; even adherents to neutralization tread carefully so that they would not be perceived as advocates of careless dairying. For instance, when Otto Hunziker spoke on cream neutralization in 1918, he vowed repeatedly that "I do not want to say or do anything that might, in any way detract from the importance of proper care of cream on the farm." Hunziker, "Neutralization of Cream for Buttermaking," 1. Debates regarding acceptable methods to purify dairy products had parallels with other manufactured foods, such as ketchup. Ann Vileisis, "Are Tomatoes Natural?" in Martin Reuss and Stephen Cutcliffe, eds., *The Illusory Boundary: Environment and Technology in History* (Charlottesville: University of Virginia Press, 2010), 224.

47. Wisconsin Buttermakers' Association, Records, 1902–1950, box 1, folder 1, Wis Mss WF, WHS, Madison, Wisc.

48. Bruce Price, "Marketing Country Creamery Butter," *University of Minnesota Agricultural Experiment Station Bulletin 244* (March 1928): 23; 1921 agreement, box 2, A. J. McGuire Papers, box 2, MNHS; H. Bruce Price, "Marketing Country Creamery Butter by a Cooperative Sales Agency," *University of Minnesota Agricultural Experiment Station Bulletin 244* (St. Paul: University of Minnesota, March 1928) 13; Larrabee and Wilster, "The Butter Industry of Oregon," 15.

49. Weekly report ending June 22, 1918, A. J. McGuire Papers, box 1, folder March–December, 1918.

50. Tax on Adulterated Butter. House Report Number 1427, January 31, 1913, 62nd Congress, Third Session (Washington, D.C.: Government Printing Office, 1913).

51. "Ruling on Butter from Decomposed Cream," *Creamery and Milk Plant Monthly* (February 1921): 23.

52. "Neutralization Permitted," *Hoard's Dairyman* 62 (September 23, 1921): 260; "Neutralizer and Adulterated Butter," *Hoard's Dairyman* 61 (February 25, 1921): 226; "Neutralized Butter Ruling," *Hoard's Dairyman* 61 (March 4, 1921): 264–265.

53. "Neutralizer Hearing in Washington," *Hoard's Dairyman* 61 (March 11, 1921): 316.

54. James Helme, "Butter Guaranteed Pure by the State," *Housewives' League Magazine* 9 (March 1917): 10.

55. Schaars, 'The Butter Industry in Wisconsin', 238–239.
56. A. J. McGuire, "A Days Work on a Dairy Farm," 4, A. J. and Marie McGuire Papers, MNHS.
57. McGuire, "A Days Work on a Dairy Farm," 4–8.
58. W. A. Henry, *Feeds and Feeding: A Hand-Book for the Student and Stockman*, 12th ed. (Madison: W. A. Henry, 1912), 249; W. A. Henry and F. B. Morrison, *Feeds and Feeding: A Handbook for the Student and Stockman*, 18th ed. (Madison, Wisc.: Henry-Morrison Company, 1922).
59. Geoff Cunfer, "Manure Matters on the Great Plains Frontier," *Journal of Interdisciplinary History* 34 (Spring 2004): 548–549, 563.
60. Gifford Diaries, 1923–31, item 7, box 1, SC 18734, NYSL.
61. Gifford Diaries, 1911, item 4, box 1, SC 18734, NYSL.
62. Rangnar and Margaret Segerstrom, interview by Dale Treleven, September 29, 1976, WAOHP, WHS, tape 2, side 1.
63. Franklin Pope Wilson to USDA Dairy Division, September 13, 1923, Franklin Pope Wilson Papers, box 1, Mss 1 W6933a, VHS.
64. Both Henry, *Feeds and Feeding* volumes; O. F. Hunziker, "Report of the Dairy Husbandry Department," *Twenty-Ninth Annual Report of the Purdue University Agricultural Experiment Station* (Lafayette, Ind.: Purdue University, 1916), 29.
65. Henry, *Feeds and Feeding*, (12th ed.) 407.
66. Henry and Morrison, *Feeds and Feeding*, (18th ed.) 384.
67. "Following the Feeder: The Proper Feeding of Cows on Pasture and the Improvement of Pasture," *Hoard's Dairyman* 77 (April 10, 1932): 191.
68. John Chamberlain, "More Silos," *Hoard's Dairyman* 31 (March 30, 1900): 142; W. C. E., "Silage Odor in Milk," *Hoard's Dairyman* 36 (February 17, 1905): 32; W. F. McSparran, "Comment on the Silo," *Hoard's Dairyman* (March 17, 1905): 154; C. O. O., "Effect of Feeding Silage," *Hoard's Dairyman* 41 (April 1, 1910): 321.
69. Albert Brodell and Thomas Kuzelka, "Harvesting the Silage Crops," *U.S. Department of Agriculture, Statistical Bulletin Number 128* (Washington, D.C.: Government Printing Office, 1953), 2, 10; H. Howard Biggar, "The Old and the New in Corn Culture," in U.S. Department of Agriculture, *Annual Yearbook of the United States* (Washington, D.C.: Government Printing Office, 1919), 134.
70. Gifford Diaries, item 5, box 1, SC 18734, New York State Manuscripts and Special Collections.
71. Henry and Morrison, *Feeds and Feeding*, 226–227.
72. A. J. McGuire, "Sweet Clover Pasture," 1929. McGuire Papers, MNHS.
73. W. G. Kaiser, "Increasing Milk and Cream Profits," *Hoard's Dairyman* 76, no. 12 (June 25, 1931): 464; A. C. Dahlberg and J. C. Marquardt, "Insulated Tanks for Cooling Milk," *Hoard's Dairyman* 76, no. 13 (July 10, 1931): 484, 501; R. S. Poulter, "Cooling Milk on the Farm," *Hoard's Dairyman* 76, no. 16 (August 25, 1931): 569, 592.
74. 1927 Land O'Lakes pamphlet, box 2, McGuire Papers, MNHS; J. R. Dawson, "Ice-Well Refrigeration for Dairy Farms Works Well at Mandan, N. Dak.," in U.S. Department of Agriculture, *Annual Yearbook of Agriculture, 1931* (Washington, D.C.: Government Printing Office, 1931), 307–310.
75. Photograph of cream cooling in Caldwell County, N.C., 1929, Record Group 16-G, box 242, National Archives, College Park, Md.
76. Price, "Marketing County Creamery Butter," 18.
77. January 1, 1904: Record of Articles and Bylaws, also Proceeding of Meetings of the Hill-Point Creamery Association, box 2, folder 1, WI Dairies Cooperative Records, Mss 473, WHS.
78. Schaars, "The Butter Industry of Wisconsin," 76; Price, "Marketing Country Creamery Butter," 22.

79. Lucia interview, tape 1, side 2; minutes of January 11, 1911, meeting, Mt. Tabor Creamery Company, box 6, folder 6, WI Dairies Cooperative, Records, Mss 473, WHS.

80. Ruble, *Men to Remember*, 42.

81. Schaars, "The Butter Industry of Wisconsin," 78.

82. "Cream Grading in Kentucky," *Creamery and Milk Plant Monthly* 12 (June 1923): 44–45; Otto Hunziker, "The Why and How of Uniformity of Butter," 5–6, speech to the Annual Meeting of the Michigan Allied Dairy Association, Kalamazoo, Michigan, February 7, 1924, Hunziker Papers, WHS.

83. Guthrie, *The Book of Butter*, 188–189.

84. Mayo, *The American Grocery Store*, 78, 84–91; Lizabeth Cohen, *Making a New Deal: Industrial Workers in Chicago, 1919–1939* (Cambridge: Cambridge University Press, 1990), 106–107; Walter Hayward and Percival White, *Chain Stores: Their Management and Operation*, 2nd ed. (New York: McGraw-Hill, 1925), 356–378.

85. "Retail Prices of Food in the United States," *Monthly Labor Bulletin* (January 1921): 62.

86. Deutsch, *Building a Housewife's Paradise*, 45.

87. See one consumer's complaint to the grocer for procuring salty butter in John Bartlett, "Personal Letters Good Practice for Business Correspondence," *Printer's Ink Monthly* 109 (1919): 170.

88. Deutsch, *Building a Housewife's Paradise*, 46–47, 51–52.

89. American Stores Company to Milltown Co-op Creamery, September 8, 1920; J. A. Simmons, American Stores Company to Mr. E. W. Shepherd, September 2, 1920, George Nelson Papers, box 1, folder Correspondence 1908–1920, Wis MSS CA, WHS.

90. Rachel Bowlby, *Carried Away: The Invention of Modern Shopping* (New York: Columbia University Press, 2001), 34–37.

91. Richard Franken and Carroll Larrabee, *Packages That Sell* (New York: Harper & Brothers, 1928), 100.

92. Paterson Parchment Paper Company, *Better Butter* (Passiac, N.J.: Paterson Parchment Paper Company, 1915), 1, Hagley Museum and Library, Trade Catalog Collection, Wilmington, Del.

93. Schaars, "The Butter Industry of Wisconsin," 222.

94. "Nature's Best Makes This *Best* Butter," Accession 59, box 28, folder 3, Butter Companies, Ayer Collection, Archives Center, NMAH.

95. Meadow Gold Advertisement, box 28, folder 3, Butter Companies, 1904–1921, Accession 59, Ayer Collection, Archives Center, NMAH.

96. Ruble, *Men to Remember*, 241.

97. Ruble, *Men to Remember*, 140.

98. Schaars, "The Butter Industry of Wisconsin," 222.

99. Meadow Gold Butter Advertisements, 1911, box 28, folder 3, Ayer Collection, Archives Center, NMAH; Fairmont Creamery Company, "A Part of Every Good Meal," Dairy, box 1, folder 28, Warshaw Collection, NMAH.

100. Aaron John Ihde and Stanley Becker, "Conflict of Concepts in Early Vitamin Studies," *Journal of the History of Biology* 4, no. 1 (Spring 1971): 34; Rima Apple, *Vitamania: Vitamins in American Culture* (New Brunswick: Rutgers University Press, 1996), 4; Michael Ackerman, "The Nutritional Enrichment of Flour and Bread," in *The Technological Fix: How People Use Technology to Create and Solve Problems*, ed. Lisa Rosner (New York: Routledge, 2004), 81; Jessica Mudry, *Measured Meals: Nutrition in America* (Albany: SUNY Press, 2009), 49.

101. Elmer V. McCollum, *A History of Nutrition: The Sequence of Ideas in Nutrition Investigations* (Boston: Houghton Mifflin, 1957), 190–191; Mudry, *Measured Meals*, 40–41.

102. Harvey Levenstein, *Revolution at the Table: The Transformation of the American Diet* (Berkeley: University of California Press, 2003), 147–160; E. V. McCollum and Nina Simmonds, *The Newer Knowledge of Nutrition: The Use of Foods for the Preservation of Vitality and Health* (New York: Macmillan, 1925) 282, 289–290; "Milk Is Our Greatest Protective Food,"

Creamery and Milk Plant Monthly 8, no. 3 (March 1919): 27; "Selling Health to the Nation: That Is What Dr. E. V. McCollum Says Is the Business of the Dairy Industry," *Creamery and Milk Plant Monthly* 16, no. 2 (February 1927): 27–30; Edith Williams, "Health Appeal in Today's Advertising," *Creamery and Milk Plant Monthly* 14, no. 6 (June 1925): 37.

103. McCollum, *A History of Nutrition,* 218.

104. Haven Emerson, "Per Capita Milk Consumption from the Point of View of the Public Health Officer," *American Journal of Public Health* 14, no. 4 (April 1924): 291–292; M. O. Maughan, Secretary of National Dairy Council, "Did It Run?" *Hoard's Dairyman* 67, no. 20 (May 30, 1924): 728; Lulu Hunt Peters, *Diet for Children (and Adults) and the Kalorie Kids* (New York: Dodd, Mead, 1924), 22.

105. Edward Meigs, "Vitamin A Value of Plant Feeds Fully Accounted for by Their Carotene Content," 324–326. Meigs argues vitamin A deficiency causes premature births in livestock. The 1939 Agricultural Yearbook, *Food and Life,* focuses on nutrition exclusively—half in people, half in livestock rations.

106. Fairmont Creamery Company Brochure, 1919, Warshaw Collection of Business Americana, Dairy, box 1, folder 28, Archives Center, NMAH.

107. Jacobs, assistant to Secretary of Agriculture to Wm. Morgan, Member of Commerce, June 28, 1941, Folder Dairying, June 28–November 26. See also John Brandt to Claude Wickard, October 28, 1941; both in box 285, RG 16, SOA, GC, 1906–76, NARA.

108. This contrasted with their position on vitamin-fortified white flour. A. J. Carlson, "What Is Wrong with the American Diet?" *Northwest Medicine* 41 (October 1941): 335–336; Ackerman, "Nutritional Enrichment of Flour and Bread," 80–82.

109. "The Comparative Nutritional Value of Butter and Oleomargarine," *Journal of the American Medical Association* 119 (August 22, 1942): 1425–1427, reprinted in House Committee on Agriculture, *Oleomargarine: Hearings on H.R. 2400,* 78th Congress, 1st sess., 1943, 15–16.

110. Reprint of Pamphlet 118, Food and Nutrition Board, National Research Council, in House Committee on Agriculture, *Oleomargarine: Hearings on H.R. 2400,* 78th Congress, 1st sess., 1943, 38.

111. National Dairy Advertisement, box 0s-351, Ayer Collection, Archives Center, NMAH.

112. Weist, *Butter Industry in the United States,* 241–242, 256–257; "Fighting Oleo for Half Century," *Hoard's Dairyman* 76, no. 4 (February 25, 1931): 129, 134, 164; Ruth Dupré, "'If It's Yellow, It Must Be Butter': Margarine Regulation in North America since 1886," *Journal of Economic History* 59 (June 1999): 353–371.

113. "Yellow Color in Milk and Cream," *Hoard's Dairyman* 67, no. 5 (February 15, 1924): 163.

114. McKay, "Facts about Butter," 34.

115. E. B. Heaton, "Cocoanut Cow, National Menace," *Hoard's Dairyman* 62 (August 26, 1921): 138; Edwy Reid, "Congress Hears Symphony on the Cocoanut Cow," *Hoard's Dairyman* 62 (August 26, 1921): 135.

116. Dairymen's Supply Company, *Our Aim "Best of Everything from the Dairy": Apparatus and Supplies for the Handling of Milk and Its Products,* Catalog 24 (Landsdowne, Penn.: Dairyman's Supply Company, 1915), 198, Hagley Museum and Library, Trade Catalog Collection.

117. "Fighting Oleo for Half Century," *Hoard's Dairyman* 76 (February 25, 1931): 134, 164; "Yellow Oleomargarine," *Hoard's Dairyman* 76 (January 10, 1931): 12.

118. "Yellow Oleo in California," *Hoard's Dairyman* 76 (February 10, 1931): 98.

119. Representative Ketcham, "The Dairy Cow Versus the Palm Tree—Another Legislative Victory for the Dairy Industry," *Congressional Record.* 71st Cong., 3d sess., 1931, vol. 7, pt. 7: 7253.

120. Representative Davis of Tennessee and Representative Menges of Pennsylvania, speaking for the Brigham Oleomargarine Bill, on February 25, 1931, to the House, H.R. 366, 71st Cong., 3rd sess., *Congressional Record* 74 (February 25, 1931): H6002, H6003.

121. The 1920 Census of Population identified 99 percent of the nation's dairy farmers to be white. Steven Ruggles, J. Trent Alexander, Katie Genadek, Ronald Goeken, Matthew B.

Schroeder, and Matthew Sobek, *Integrated Public Use Microdata Series: Version 5.0* (Minneapolis: Minnesota Population Center, 2010). But the census data is highly problematic, for the majority of the farm people engaged in dairying were general farmers, especially in the cream trade. Likely, only the highly capitalized, fluid-milk farms are probably classified as "dairy farms," and so the number of people engaged in dairying, including families of color, were simply listed as "farmers." Only the 1910 and 1920 population census included the occupational category dairy farmer, so it is impossible to trace this category over time. The agricultural census has no clear category for "dairy farmer," only the category "farms with dairy cows."

122. Martha Crampton Howard, "The Margarine Industry in the United States: Its Development under Legislative Control" (Ph.D. diss., University of Wisconsin-Madison, 1951), 291, 317.

123. Judith Russell and Renee Fantin, *Studies in Food Rationing*, War Administration: Office of Price Administration, General Publication 13 (Washington, D.C.: Government Printing Office, 1947), 185.

124. S. Louise Grant Papers, quoted in *Produce and Conserve, Share and Play Square: The Grocer and the Consumer on the Home-Front Battlefield during World War II*, ed. Barbara McLean Ward (Portsmouth, N.H.: Strawbery Banke Museum, 1994), 192.

125. Records of the Office of the Secretary of Agriculture, General Correspondence, 1906–1976, box 286, folder Dairying 1 Dairy Products April 1–July 1, 1943, RG 16, SOA, GC, 1906–70, NARA.

126. S. Louise Grant Papers, box 1, series I: Sarah Louise Grant 1909–1990, MS 91, Thayer Cumings Library and Archives Collection, Strawbery Banke Museum, Portsmouth, N.H.; Russell and Fantin, *Studies in Food Rationing*, 201; Howard, "The Margarine Industry in the United States," 316–319.

127. *Oleomargarine: Hearings on H.R. 2400*, 78th Congress, 1st sess., 1943, 26 (Poage) and 74–75 (LaGuardia); Fred Wallace to Secretary of Agriculture, "Liquidation of Dairy Herds in California," December 10, 1942, box 632, RG 16, SOA, GC, 1906–75, NARA.

128. Louis Stradling to Ezra T. Benson, February 1954, box 2395, folder Dairy Products, January 1–April 12, RG 16, SOA, GC, 1906–76, NARA.

Chapter 3

1. John Michels, "Wisconsin Buttermakers' Meeting," *Hoard's Dairyman* 36 (March 10, 1905): 123; M. Mortensen and J. B. Davidson, "Creamery Organization and Construction: Part 1," *Agricultural Experiment Station Bulletin 139* (Ames: Iowa State University, 1913), 128; Donald Carmichael, "Forty Years of Water Pollution Control in Wisconsin: A Case Study," *Wisconsin Law Review* (Spring 1967): 418–419.

2. Otto Hunziker, *The Butter Industry*, 3rd ed. (Lagrange, Ill.: Hunziker, 1940), 30.

3. Paul Sutter, *Driven Wild: How the Fight against Automobiles Launched the Modern Wilderness Movement* (Seattle: University of Washington Press, 2002), 24–27; Warren Belasco, *Americans on the Road: From Autocamp to Motel, 1910–1945* (Cambridge: MIT Press, 1979), 30–39.

4. Conflict between those who wanted to use nature for tourism and those who wanted to use it as a site for rural industry was not unique to dairy manufacturing districts. See, for instance, the conflict between cannery owners and hotel owners on California's coast in Connie Chiang, *Shaping the Shoreline: Fisheries and Tourism on the Monterey Coast* (Seattle: University of Washington Press, 2008), 91–101.

5. Unattributed speech, "Early Thoughts on Improvement" (1928–1931), box 1, folder 3, CBC Records, WHS.

6. Linda Nash, *Inescapable Ecologies: A History of Environment, Disease, and Knowledge* (Berkeley: University of California Press, 2006), 84–85.

7. U.S. Department of Agriculture, *Annual Yearbook of Agriculture, 1931*, 15–16, 24–25; U.S. Department of Agriculture, *Annual Yearbook of Agriculture, 1927*, 111–114; for milk prices: Dairy and Poultry Statistics, U.S. Department of Agriculture, *Annual Yearbook of Agriculture* (Washington, D.C.: Government Printing Office, 1931), 914, and U.S. Department of Agriculture, *Annual Yearbook of Agriculture, 1935* (Washington, D.C.: Government Printing Office, 1935), 607; for butter: U.S. Department of Agriculture, *Annual Yearbook of Agriculture, 1931*, 921; John D. Black, *The Dairy Industry and the AAA* (Washington, D.C.: Brookings Institution, 1935), 70–71.

8. *Fisher v. Zumwalt*, 128 Cal. 493; *Behnisch v. Cedarburg Dairy Company*, 180 Wis 34; *Alex Ruthven v. Farmers Cooperative Creamery Company*, 140 Iowa 570; *W. T. Perry v. Howe Cooperative Creamery Company*, 125 Iowa 415; *Johnson v. Kraft-Phenix Cheese Corporation*, 19 Tenn. App. 648.

9. Bureau of Sanitary Engineering, Wisconsin State Board of Health, *Progress Report of the State Committee on Water Pollution* (Madison: Wisconsin State Board of Health, 1931), Table IV—Industrial Waste Investigations, 57–114.

10. For more on defining waste from want, see Susan Strasser, *Waste and Want: A Social History of Trash* (New York: Metropolitan, 1999), 5–9. For other manufacturers' efforts to reuse waste, see Samuel P. Hays, *Conservation and the Gospel of Efficiency: The Progressive Conservation Movement, 1890–1920* (Cambridge: Harvard University Press, 1989), 123; Craig Colten and Peter Skinner, *The Road to Love Canal: Managing Waste before EPA*, (Austin: University of Texas Press, 1996), 143. In other industries, World War II tended to diminish efforts to find new uses for waste, rather than further this process.

11. For the history of chemurgy, see Caroll Pursell Jr., "The Farm Chemurgic Council and the United States Department of Agriculture, 1935–1939," *Isis* 60 (Autumn 1969): 307–317; David Wright, "Alcohol Wrecks a Marriage: The Farm Chemurgic Movement and the USDA in the Alcohol Fuels Campaign in the Spring of 1933," *Agricultural History* 67 (Winter 1993): 36–66; Randall Beeman, "'Chemivisions': The Forgotten Promises of the Chemurgy Movement," *Agricultural History* 68 (1994): 23–45; David Wright, "Agricultural Editors Wheeler McMillen and Clifford V. Gregory and the Farm Chemurgic Movement," *Agricultural History* 69 (Spring 1995): 272–287; Anne Effland, "'New Riches from the Soil': The Chemurgic Ideas of Wheeler McMillen," *Agricultural History* 69 (Spring 1995): 288–297; Mark Finlay, "The Failure of Chemurgy in the Depression-Era South," *Georgia Historical Quarterly* 81 (Spring 1997): 78–102; William J. Hale, *The Farm Chemurgic: Farmward the Star of Destiny Lights Our Way* (Boston: Stratford Company, 1934), ii.

12. Christy Borth, *Pioneers of Plenty: The Story of Chemurgy* (Indianapolis: Bobbs-Merrill, 1939), 22–23.

13. Wheeler McMillen, *New Riches from the Soil: The Progress of Chemurgy* (New York: D. Van Nostrand, 1946), 6, 22, 32–35; Borth, *Pioneers of Plenty*, 24–29.

14. Lynn Dumenil, *The Modern Temper: American Culture and Society in the 1920s* (New York: Hill and Wang, 1995), 144–149; William Leuchtenburg, *The Perils of Prosperity, 1914–1932*, 2nd ed. (Chicago: University of Chicago Press, 1993), 225–229.

15. Henry Henderson and David Woolner, eds., *FDR and the Environment* (New York: Palgrave-Macmillan, 2005); Beeman, "'Chemivisions': The Forgotten Promises of the Chemurgy Movement," 41; Hale, *The Farm Chemurgic*, 7–15.

16. Sarah Phillips, *This Land, This Nation: Conservation, Rural America, and the New Deal* (New York: Cambridge University Press, 2007), 9–11.

17. Donald Worster, *Dust Bowl: The Southern Plains in the 1930s* (New York: Oxford University Press, 1979); Neil Maher, *Nature's New Deal* (New York: Oxford University Press, 2008): Neil Maher, "A New Deal Body Politic: Landscape, Labor, and the Civilian Conservation Corps," *Environmental History* 7, no. 3 (July 2002): 435–461.

18. Henry E. Alvord, "Modern Dairy," *Current Literature* 29, no. 5 (November 1900): 598.

19. Edwin Sutermeister, *Casein and Its Industrial Applications* (New York: Chemical Catalog Company, 1927), 84.

20. Borth, *Pioneers of Plenty*, 13–15; "Synthetic Fabrics in Fashion Parade," *New York Times*, December 8, 1939, 18; "New Casein Fiber Due for Wide Use," *New York Times*, November 9, 1941; "Casein Fabrics Displayed Here in Gowns Worn by Stage Stars," *New York Times*, November 25, 1941; Edwin Sutermeister and Frederick Browne, *Casein and Its Industrial Applications* (New York: Reinhold, 1939), 398.

21. "Casein, Milk Protein, Has Diversified Uses," *News Bulletin of the National Farm Chemurgic Council* 2, no. 22 (November 25, 1941): 169.

22. Margaret Segerstrom, interview by Dale E. Treleven, September 29, 1976, WAOHP, WHS.

23. "The Whole Milk Creamery Movement: What Is It? What Does It Mean? Where Will It Lead?" *National Butter and Cheese Journal* 28, no. 1 (January 12, 1937): 10, 12, 14–15.

24. Floyd Lucia, interview with Dale Treleven, WHS.

25. O. E. Reed, "Food Value of Skim Milk," *Hoard's Dairyman* 68 (September 26, 1924): 270.

26. National Farm Chemurgic Council, "New Uses for Milk Byproducts—Part 2," *News Bulletin of the National Farm Chemurgic Council* 1, no. 2 (January 25, 1940): 12; "United States Still Imports More Than Half of Its Casein Supply," *Glass Lining* 4, no. 5 (October 29, 1929): 16; untitled article, *Glass Lining* 4, no. 2 (May 1928): 15; Sutermeister and Browne, *Casein and Its Industrial Applications*, 400.

27. U.S. Department of Agriculture, *Annual Yearbook of Agriculture, 1931*, 72.

28. Sutermeister and Browne, *Casein and Its Industrial Applications*, 399–400.

29. Sutermeister and Browne, *Casein and Its Industrial Applications*, 135.

30. For "apostles of modernity," see Roland Marchand, *Advertising the American Dream: Making Way for Modernity, 1920–1940* (Berkeley: University of California Press, 1985), 1.

31. National Farm Chemurgic Council, "New Uses for Milk Byproducts—Part 2," *News Bulletin of the National Farm Chemurgic Council* 1, no. 2 (January 25, 1940): 12.

32. National Farm Chemurgic Council, "Casein Paint," *News Bulletin of the National Farm Chemurgic Council* 1, no. 4 (February 25, 1940): 28–29; DuPont de Nemours and Company, "DuPont Casein-Lithopone Flat Wall Finish," 1940, Hagley Museum and Library, Trade Catalog Collection, D93, 1940a; National Farm Chemurgic Council, "New Uses for Milk Byproducts,—Part 2," 11.

33. George Brother, "Casein Plastics," in *Casein and Its Industrial Applications* (1927), 142–165, 185–214; P. Ch. Christensen, "Colored Casein Buttons Can Now Be Produced Quickly from White Stocks," *Plastics: Periodical Devoted to the Manufacture and Use of Composition Products* 6, no. 2 (February 1930): 116, 121; Stephen Bass, *Plastics and You* (N.p.: Eastwood-Steli Company, 1947): 95–96.

34. Waldemar Kaempffert, "This Week in Science: Wool Made from Skim Milk," *New York Times*, May 10, 1936.

35. "New Wool-Like Fiber Is Yielded by Casein," *New York Times*, August 15, 1938.

36. Wheeler McMillen, *New Riches from the Soil: The Progress of Chemurgy* (New York: D. Van Nostrand, 1946), 251–252; "New Casein Fiber Due for Wide Use," *New York Times*, November 9, 1941.

37. National Farm Chemurgic Council, "Ohio Milk Wool," *News Bulletin of the National Farm Chemurgic Council* 1, no. 11 (June 10, 1940): 91–92.

38. "Casein Fiber Plant for U.S. Planned," *New York Times*, November 24, 1937; "New Wool-like Fiber Is Yielded by Casein," *New York Times*, August 15, 1938; "Synthetic Fabrics in Fashion Parade," *New York Times*, December 8, 1939; "Casein Fabrics Displayed Here in Gowns Worn by Stage Stars," *New York Times*, November 25, 1941, p. 28; Martha Parker, "The Beauty Quest: A 'Milk-Fed Permanent Wave,'" *New York Times*, February 27, 1942; "Forest and Farm Contribute to Garment Fabrics of the Future," *New York Times*, October 8, 1942; Herbert Koshetz, "See Protein Fibers Due for Wider Use," *New York Times*, April 28, 1946.

39. Bass, *Plastics and You*, 97–98. Wool-growers, perhaps unsurprisingly, were not enthusiastic about casein cloth's potential to replace its products. See "Laboratory Notes," *New York Times*, May 1, 1938.

40. "Casein Fiber Plant Sold," *New York Times*, February 9, 1948.

41. National Farm Chemurgic Council, "Milk Casein for Textiles," *News Bulletin* 2, no. 21 (November 10, 1941), 160.

42. A. O. Dahlberg, "The Manufacture of Casein," in *Casein and Its Industrial Applications*, 60.

43. Borth, *Pioneers of Plenty*, 264.

44. J. S. Abbott, "The Food Value and Economics of Skim Milk," *American Journal of Public Health* 30, no. 3 (March 1940): 238.

45. McMillen, *New Riches from the Soil*, 6, 10, 14.

46. Nestle's Milk advertisement, Milk, box 2, Warshaw Collection of Business Americana, 1850–1950, Archives Center, NMAH.

47. *Proceedings of the Eighth Annual Meeting of the American Dried Milk Institute* (Chicago: American Dried Milk Institute, 1933), 23.

48. For skim milk drying processes, see Merrell-Soule Company, *Merrell-Soule Products: Powdered Milk and None Such Mince Meat* (Syracuse, N.Y.: Merrell-Soule Company, 1919), 14–17; George Holm, "Dried Skim Milk Added to Other Foods Improves Their Nutritive Value," in U.S. Department of Agriculture, *Annual Yearbook of Agriculture, 1935*, (Washington, D.C.: Government Printing Office, 1936), 171–172; Floyd Lucia interview, tape 2, side 1; for Wisconsin Dried Milk Pool, see Floyd Lucia, October 5, 1937 Speech to the Wisconsin Buttermakers' Association, box 1, folder 2, Wisconsin Buttermakers' Association Records, 1902–1950, WHS.

49. "Dried Skim Milk Added to Other Foods Improves Their Nutritive Value," 171–172.

50. Louise Stanley, "Address," in *Ninth Annual Meeting of the American Dry Milk Institute*, April 18–19, 1934 (Chicago: American Dry Milk Institute, 1934), 123, 126.

51. Chatfield, "Nutritional Needs of the South," in *Proceedings of the Sixth Annual Meeting, American Dry Milk Institute* (Chicago: American Dried Milk Institute, 1931), 42–49; Louise Stanley, "Consumer Acceptance of Dry Milk Solids," in *Proceedings of the Eleventh Annual Meeting of the American Dry Milk Institute* (Chicago: American Dry Milk Institute, 1936), 54–60; Janet Poppendieck, *Breadlines Knee-Deep in Wheat: Food Assistance in the Great Depression* (New Brunswick, N.J.: Rutgers University Press, 1986), 138–139.

52. Hugh Cook and George Day, *The Dry Milk Industry: An Aid in the Utilization of the Food Constituents of Milk* (Chicago: American Dry Milk Institute, 1947), 102.

53. See speeches by O. T. Goodwin, R. J. Howat, and L. W. Nolte on the Michigan Feed Sales Campaign, in *Proceedings of the Tenth Annual Meeting of the American Dry Milk Institute* (Chicago: American Dry Milk Institute, 1935), 2–30; "Chick Starter, Growing Mash, Broiler Mash," *Feed Service Bulletin 210* (1936): 15; C. W. Sievert, "Feeding for Hatchability," *Feed Service Bulletin 211* (1936): 3–10; E. N. Craig, "Market Development," in *Proceedings of the Fifteenth Meeting of the American Dry Milk Institute*, April 17, 1940 (Chicago: American Dry Milk Institute, 1940), 5–6.

54. Clyde Beardslee, "Dairy Problems under War Time Conditions," in *Proceedings of the Seventeenth Meeting of the American Dry Milk Institute* (Chicago: American Dry Milk Institute, April 23–24, 1942), 72; C. E. Gray, "Striking the Keynote: Quality," in *Proceedings of the Seventeenth Meeting of the American Dry Milk Institute*, 1.

55. The Food and Nutrition Board's members included chemists and nutritionists who assessed the qualities of foodstuffs and their physiological effects on human development. A second National Research Council group, the Committee on Food Habits, headed by Margaret Mead, was focused on the cultural barriers and practices that contributed to food choices. Amy Bentley, *Eating for Victory: Food Rationing and the Politics of Domesticity* (Urbana: University of Illinois Press, 1998), 24–25.

56. W. C. Rose et al., "Nation's Protein Supply," 8, Records of the National Research Council, Food and Nutrition Board, 1942, Committee on Milk, Meat, and Legumes, Archives of the National Academy, Washington, D.C.

57. Harvey Levenstein, *Revolution at the Table: The Transformation of the American Diet* (Berkeley: University of California Press, 2003), 138–139; O. E. Reed, "Changing from Cream to Whole Milk Deliveries," in *Proceedings of the Eighteenth Annual Meeting of the American Dry Milk Institute* (Chicago: American Dry Milk Institute, April 14–15, 1943), 49. Cottage cheese played a big role in the World War I meatless effort. For examples of the food administration's work, see "The United States' Best Recipes: Dishes with the Experts of the Food Administration Have Worked Out," *Ladies Home Journal* 35, no. 6 (June 1918): 37.

58. United Dairy Equipment Company, *Sales Manual and Instructions for Installation and Operation of the World-Famous Mechanical Cow* (West Chester, Penn.: United Dairy Equipment Company, 1943), 3, 58, Hagley Museum and Library, Trade Catalog Collection.

59. Major C. J. Herman, "Dry Milk Solids and Its Place in the Army Ration," in *Proceedings of the Seventeenth Annual Meeting of American Dry Milk Institute*, 68–70.

60. "Badger's War Quality Program," 1942 Annual Report, Consolidated Badger Cooperative Records, box 1, folder 12, Green Bay Mss 104. WHS.

61. Raymond Kern to Rochester Cooperative Dairy—reprinted in address by W. W. Thompson, in *Proceedings of the Eighteenth Annual Meeting of the American Dry Milk Institute*, 53.

62. John Summe to Claude Wickard, November 12, 1942, Folder Dairy-Dairying Products, January 1–March 31, 1943, box 837, RG 16, SOA, GC, 1906–70, NARA; for the history of World War II scrap drives, see Strasser, *Waste and Want*, 226–263.

63. U.S. Department of Agriculture, Bureau of Agricultural Economics, *Production of Manufactured Dairy Products, 1946* (Washington, D.C.: Government Printing Office, May 1948), 7.

64. Hugh Cook and George Day, *The Dry Milk Industry: An Aid in the Utilization of the Food Constituents of Milk* (Chicago: American Dry Milk Institute, 1947), 3.

65. R. M. Bethke, "Feeding Pigs, Chickens, and Calves," and B. W. Fairbanks, "Discussion of Feeding," in *Proceedings of the Eighteenth Annual Meeting of the American Dry Milk Institute*, 27–31; "Milk Alternates, Replacements, Substitutes," in *Proceedings of the Nineteenth Annual Meeting of the American Dry Milk Institute*, 2–12.

66. Margaret Mead to Frank Gunderson, March 23, 1942, Papers of the Committee on Food Habits, Division of Anthropology and Psychology, National Research Council, folder Anthropology and Psychology/Committee on Food Habits-Cooperation with Division of Biology and Agriculture, Archives of the National Academies.

67. C. M. Peterson, "The Farmer Produces, The Consumer Needs, We Must Produce ALL OF IT!" in *Proceedings of the Eighteenth Annual Meeting of the American Dry Milk Institute*, 16.

68. Bryan Blalock, "Report on H.R. 149," *Proceedings of the Eighteenth Annual Meeting of the American Dry Milk Institute*, 18–19; Bryan Blalock, "Report on H.R. 149," *Proceedings of the Nineteenth Annual Meeting of the American Dry Milk Institute*, 33–37; "Now It's Official," *Hoard's Dairyman* 89, no. 7 (April 10, 1944): 208.

69. Ralph Krauss, "Need for a Program to Effect Wider Use of Skimmilk Solids for Human Food," December 1, 1945, National Academies of Science, Food and Nutritional Board, Committee on Milk, Archives of the National Academies; Paul Stark, Director, Food Distribution Programs Branch, to Ralph Trigg, Deputy Administrator, Production and Marketing Administration—February 27, 1947, RG 16, box 1441, folder Dairy Products 6 Price Supports (1 of 2), National Archives; Hugh Cook et al., *Nonfat Dry Milk Solids: Package Preferences, Household Buying and Storage Practices in Selected Southern Areas* (Madison, Wisc.: Department of Agricultural Economics, 1954), 2; Cook, *The Dry Milk Industry*, 143; Margaret Morris, "School Lunches," in *Proceedings of the Twenty-Second Annual Meeting of the American Dry Milk Institute* (Chicago: American Dried Milk Institute, April 22–23, 1947), 29–31.

70. W. T. Hutchinson, Assistant Secretary of Agriculture to Hon. Alexander Wiley, September 13, 1949, folder 3, Dairy Products, July 1 (nd), box 1701, RG 16, SOA, GC, 1906–70, NARA.

71. Poppendieck, *Breadlines Knee-Deep in Wheat*, 242.

72. Schultz to Merlin Hull, March 19, 1947, folder Dairy Products 6 Price Support (1 of 2)—January 1–May 13, box 1441, RG 16, SOA, GC, 1906–70, NARA.

73. A. J. Lowland to Senator Alexander Wiley, January 1949, folder Dairy Products—January 1–June 30, box 1701, RG 16, SOA, GC, 1906–70, NARA.

74. Charles Brannan to Charles Holman, Secretary of National Milk Producers' Federation, June 6, 1952, folder Dairy Products 6—School Lunch Program, box 2073, RG 16, SOA GC 1906–70, NARA.

75. Grover Turnbow, President of Foremost Dairies Incorporated, to Mr. N. R. Clark, U.S. Department of Agriculture, July 16, 1954, folder Dairy Products June 8–August 5, box 2395. RG 16, SOA, GC, 1906–70, NARA.

76. D. B. Hand, "Present Status and Trends in the Utilization of Non-Fat Dry Milk Solids for Human Consumption," November 14, 1952, entry 6, box 2, folder Associations and Communications, 1952, Records of the Bureau of Dairy Industry, 1944–1953, Record Group 152, NARA.

77. E. I. DuPont de Nemours and Company, *Design for Selling: A Study of Impulse Buying* (Wilmington, Del.: E. I. DuPont de Nemours and Company, 1946), 6, 8–9, Hagley Museum and Library Pamphlet Collection.

78. Twenty-Second Annual Report of Consolidated Badger Cooperative, 1953; Folder 12—Annual Reports, 1931–54, box 1, CBC Papers, WHS. Land O'Lakes also emphasized its dry milk in supermarkets. Jack El-Hai, *Celebrating Tradition: Building the Future: Seventy-Five Years of Land O'Lakes* (Minneapolis, Minn.: Land O'Lakes, 1996), 74.

79. Hand, "Present Status and Trends," 3, November 14, 1952; entry 6, box 2, folder Associations and Communications, 1952, BDI, 1944–1953, RG 152, NARA.

80. *Life* magazine's iconic photograph of a family in a fallout shelter, for instance, depicts a big box of the Borden's powdered milk product "Starlac" on the shelf closest to the woman pictured, suggesting its importance. The photo is on the cover of Elaine Tyler May, *Homeward Bound: American Families in the Cold War Era*, rev. ed. (New York: Basic, 1999).

81. Harvey Levenstein, *Paradox of Plenty: A Social History of Eating in Modern America* (Berkeley: University of California Press, 2003), 109.

82. Hot dogs became increasingly popular, especially among families in the postwar era. Roger Horowitz, *Putting Meat on the American Table: Taste, Technology, Transformation* (Baltimore: Johns Hopkins University Press, 2006), 96–99.

83. American Dry Milk Institute, *1953 Census of Dry Milk Distribution and Production Trends* (Chicago: American Dried Milk Institute, 1954), 8, 4, 10–11; Cook, *The Dry Milk Industry*, 97–98, 100.

84. O. E. Reed to Thomas Stephens, Special Counsel to the President, May 28, 1953, folder Cheese, box 5, entry 6, RG 152; O. E. Reed to Samuel R. Guard, March 10, 1953, folder Milk and Cream Information, entry 6, box 7; both in BDI, 1944–1953, RG 152. R. T. Hutchinson to Secretary of Defense, February 1, 1951, box 1964; George Mehren to John Byrnes, February 24, 1965, folder Dairy Products, January 1–June 4, box 4274; both in RG 16, SOA, GC, 1906–70, NARA.

85. N. W. Ayer Collection, folder National Dairy—Sheffield Farms Company Inc.—Sealtest Dairy Products 1951–1957 #2, box OS-355, Ayer Collection, Archives Center, NMAH; on byproducts in sausage, see Horowitz, *Putting Meat on the American Table*, 82, 86; A. H. Johnson, "Research Builds the Dairy Industry," in *Proceedings of the Nineteenth Annual Meeting of the American Dried Milk Institute*, 28.

86. Wilbur Foust, "Let's Sell More Fat Free Milk," *Milk Dealer* 38, no. 5 (October 1948): 80.

87. Foust, "Let's Sell More Fat Free Milk," 80; John Lancey, "Promoting Non-fat Fortified Milk," *Milk Dealer* 38, no. 5 (February 1949): 42.

88. Roberta Seid, *Never Too Thin: Why Women Are at War with Their Bodies* (New York: Prentice Hall, 1989), 105.

89. Folder Clover Dairy, Sealtest, Western MD Dairy: 1936–1937, 1957; box OS 353, folder National Dairy, 1948–1950.

90. The National Grange, *Barriers to Increased Consumption of Fluid Milk* (Washington, D.C.: Guinea Company, 1955), 19–21, Folder Dairy Products January 1–February 18, 1958, box 2558, RG 16, SOA, GC, 1906–70, NARA.

91. Vitex Laboratory advertisement, *Milk Dealer* 39, no. 6 (March 1950): 134.

92. N. W. Ayer Advertising Collection, box OS-356, folder 1953–1956—Sealtest Cottage Cheese and Milk, folder 1, Ayer Collection, Archives Center, NMAH.

93. Judson Mason, economist of National Milk Producers Association, Address at Annual Meeting of Consolidated Badger Cooperative, 1960, folder 14, Annual Meeting, 1960, box 3, CBC Records, WHS.

94. H. F. Stevens to Mallary, February 27, 1939, Mallary Farm Records, 1935–1988, Document 495, MS Acc No 2002.11, Vermont Historical Society, Barre, Vt. (hereafter VT HS).

95. H. F. Stevens to Mallary, February 27, 1939, Mallary Farm Records, VT HS.

96. Beall Farm Record Books, 1940, 1949–1950, 1954–1955, Apple River, Illinois, item in possession of the author.

97. Virginia Hollerith to Mr. Bristow, January 27, 1941. Folder 20, Correspondence 1941, Box 1, Hollerith Family papers, 1904–1985, Mss IH 7235b, Virginia Historical Society, Richmond, VA. (hereafter Hollerith Family Papers, VHS).

98. For nineteenth-century examples of scientific development in cattle breeding, see Harriet Ritvo, "The Sincerest Form of Flattery," in *Animals in Human Histories: The Mirror of Nature and Culture*, ed. Mary Henninger-Voss, 295–315 (Rochester, N.Y.: University of Rochester, 2002); Barbara Orland, "Turbo-Cows: Producing a Competitive Animal in the Nineteenth and Early Twentieth Centuries," in *Industrializing Organisms: Introducing Evolutionary History*, ed. Susan Schrepfer and Philip Scranton, 167–189 (New York: Routledge, 1994). For breeding in agriculture more largely, see Fitzgerald, *The Business of Breeding*; Jack Ralph Kloppenburg Jr., *First the Seed: The Political Economy of Plant Biotechnology, 1492–2000* (New York: Cambridge University Press, 1988); William Boyd, "Making Meat: Science, Technology, and American Poultry Production," *Technology and Culture* 42 (2001): 631–664.

99. G. C. Humphrey, "Community Breeders' Associations for Dairy Cattle Improvement," *Agricultural Experiment Station Bulletin* 189 (Madison: University of Wisconsin, 1910): 1–21; "The Tester's Column," *Hoard's Dairyman* 77, no. 7 (April 10, 1932): 194.

100. R. R. Graves, "Superior Germ Plasm in Dairy Herds," U.S. Department of Agriculture, *Annual Yearbook of Agriculture, 1936* (Washington, D.C.: Government Printing Office, 1936), 998; V. A. Rice, "The Next Best Thing to a Proved Sire," *Hoard's Dairyman* 77, no. 3 (February 10, 1932): 61, 83; A. C. Baltzer, "Which Sire Do You Want to Select?" *Hoard's Dairyman* 83, no. 5 (March 10, 1938): 139.

101. O. T. Rice, Oakton Virginia to Mr. H. E. Hutchinson, September 8, 1932, folder 11, Correspondence 1932, box 1, Hollerith Family Papers, VHS.

102. August 10, August 13, 1922, entries, Elihu Gifford Diaries, item 6, box 1, SC 18734, New York State Archives and Manuscripts, Albany, N.Y. (hereafter Gifford Diaries, NYSL).

103. March 24 entry, Boyd Sherwood Diaries, item 5, box 1, SC 22581, New York State Library Manuscripts and Special Collections, Albany, N.Y. (hereafter Sherwood Diaries, NYSL).

104. Harvey Dueholm, interviewed by Dale Treleven, December 19, 1978, tape 12, side 1, WAOHP, WHS.

105. Boyd Sherwood, October 5, 1940, diary entry, item 12, box 2, Sherwood Diaries, NYSL.

106. Perry, *Artificial Insemination of Farm Animals*, 4; Randall Swanson, "The Dairy Bull—Dr. Jekyll and Mr. Hyde," *Hoard's Dairyman* 89, no. 21 (November 10, 1944): 598; clippings in folder 1, box 1, Hollerith Family Papers, VHS.

107. E. R. Zook, "Replies to Klemm," *Hoard's Dairyman* 83, no. 4 (February 25, 1938): 109.

108. "Bull Management," *Hoard's Dairyman* 86, no. 8 (April 25, 1941): 280; H. A. Herman and F. W. Madden, *The Artificial Insemination of Dairy Cattle: A Handbook and Laboratory Manual* (Columbia, Mo.: Lucas Brothers, 1947), 72–76.

109. Perry and Bartlett, "Artificial Insemination of Dairy Cows," 5; Herman and Madden, *The Artificial Insemination of Dairy Cattle*, 12–14.

110. Ralph Phillips, "Artificial Breeding," U.S. Department of Agriculture, *Annual Yearbook of Agriculture, 1943–1947*, 113.

111. G.vL. Cole, "Artificial Insemination," *Hoard's Dairyman* 86, no. 1 (January 10, 1941): 18.

112. Cole, "Artificial Insemination," 18.

113. Pou, "A Study of Wisconsin Cooperative Artificial Insemination Associations," 50; Thomas et al., *Dairy Farming in the South*, 91–93; Bennie Brown, "Bull 'Tame as a Dog,'" *Hoard's Dairyman* 86, no. 16 (August 25, 1941): 501, 517.

114. H. A. Herman, *Improving Cattle by the Millions: NAAB and the Development and Worldwide Application of Artificial Insemination* (Columbia: University of Missouri, 1981), 7, 37; Perry, *The Artificial Insemination of Farm Animals*, 7; John William Pou, "A Study of Wisconsin Cooperative Artificial Insemination Associations" (Master's thesis, University of Wisconsin-Madison, 1947): 10–11; H. A. Herman, "Making Artificial Breeding Succeed," *Hoard's Dairyman* 95, no. 9 (May 10, 1950): 357.

115. "Information on Farm of Mr. and Mrs. R. DeWitt Mallary, Fairlee, Vermont, 1937," folder 33, General Correspondence 1937, and R. Dewitt Mallary to Selective Service Board #1, December 25, 1945, Doc 495–38, Mallary Farm Records, 1936–1974, VT HS.

116. Pou, "A Study of Wisconsin Cooperative Artificial Insemination Association," 56; DeGraff, "Making Artificial Breeding Work," 230.

117. Cole, "Artificial Insemination," 18; "Greater Breeding Efficiency," *Hoard's Dairyman* 89, no. 19 (October 10, 1944): 534; Pou, "A Study of Wisconsin Cooperative Artificial Insemination Associations," 56; E.vJ. Perry and J.vW. Bartlett, "Artificial Insemination of Dairy Cows," *Extension Bulletin 284* (New Brunswick, N.J.: Extension Service, College of Agriculture, 1955), 3.

118. By the time Mr. and Mrs. Mallary purchased their farm, Dr. Williams had graduated from Albany Medical College and began to practice medicine in Springfield, Massachusetts, the same town in which Mr. Mallary practiced law. Folder 30, Walter W. Williams Correspondence 1939–1940, Mallary Farm Records, VT HS.

119. A. H. DeGraff, "Making Artificial Breeding Work," 253; E. J. Perry, "Artificial Breeding Costs," *Hoard's Dairyman* 86, no. 17 (September 10, 1941): 533.

120. Frank Hill, interview with Lu Ann Jones, May 4, 1987, Southern Agriculture Oral History Project, Accession 773, box 18, folder 10, Archives Center, National Museum of American History, Washington, D.C.

121. Pou, "A Study of Wisconsin Cooperative Artificial Insemination Associations," 54.

122. Dewitt Mallary to Dr. Williams, March 5, 1940, folder 30; John Ellis to Mrs. Mallary, April 3, 1954; both in Mallary Farm Records, VT HS.

123. Enos Perry, "Artificial Breeding of Dairy Cows Review," *Hoard's Dairyman* 89, no. 23 (December 10, 1944): 656; Phillips, "Artificial Breeding," 117.

124. E. A. Woelffer, "How to Detect Heats," *Hoard's Dairyman* 97, no. 3 (February 10, 1952): 155.

125. Perry, "Artificial Breeding of Dairy Cows Review," 672.

126. Deborah Fitzgerald, "Farmers Deskilled: Hybrid Corn and Farmers' Work," *Technology and Culture* (1993): 324–325.

127. David Bartlett, Kenneth Moist, and Francis Spurrell, "The *Trichomonas foetus* Infected Bull," reprinted in *Proceedings of Fifth Annual Convention of the National Association of Artificial Breeders* (Springfield, Mo.: National Association of Artificial Breeders, 1952), 75–77.

128. Ernest Mercier and G. W. Salisbury, "Fertility Level in Artificial Breeding Associated with Season, Hours of Daylight, and the Age of Cattle," *Journal of Dairy Science* 30, no. 11 (November 1948): 824.

129. Bulls used for artificial insemination provide a provocative example of an organismal technology, as described in Edmund Russell's "The Garden in the Machine: Toward an Evolutionary History of Technology," in *Industrializing Organisms: Introducing Evolutionary History*, 4–5.

130. Perry, *Artificial Insemination of Farm Animals*, 11.

131. Mercier and Salisbury, "Fertility Level in Artificial Breeding," 824.

132. H. T. Stevens to Mrs. Mallary, January 24, 1942, Business Organization, folder 1, box 1, Mallary Farm Records, VT HS.

133. Herman, *Improving Cattle by the Millions*, 49, 85–87.

134. Folder 26, Frozen Semen Transfers, 1957–1970, box 498, Mallary Farm Records, VT HS.

135. Lyle Jackson and J. W. Pire, "Bulls in Pasture Together," *Hoard's Dairyman* 95, no. 3 (February 10, 1950): 102–103.

136. Mallary (Mr. or Mrs. Unknown) to Mr. M. B. Nichols, April 24, 1944, folder 21, box 496, Mallary Farm Papers, VT HS.

137. William Cronon, *Nature's Metropolis: Chicago and the Great West* (New York: Norton, 1991), 146–147.

138. Subcommittee on Dairy Waste Disposal, Dairy Industry Committee, in Cooperation with the National Technical Task Committee on Industrial Wastes, *An Industrial Waste Guide to the Milk Processing Industry* (Washington, D.C.: U.S. Department of Health, Education, and Welfare, 1953), 5–6.

139. James Joseph Flannery, "Water Pollution Control: Development of State and National Policy" (Ph.D. diss., University of Wisconsin, 1956), 53–75; "Hits at Cheese Pollution Plea," *Milwaukee Journal*, October 28, 1948; Wisconsin Cheese Makers' Association, *Proceedings of the Wisconsin Cheese Makers' Association Fifty-Sixth Annual Meeting*, November 3–4, 1947, Auditorium and Schroeder Hotel, Milwaukee, Wisconsin (Madison, Wisc.: Cantwell Press, 1913/1915–1954), 35, accessed August 8, 2006, at www.digital.library. wisc.edu/1711.dl/WI.WCA.1947; Wisconsin Cheese Makers' Association, *Proceedings of Wisconsin Cheese Makers' Association Fifty-Seventh Annual Meeting and Centennial Convention*, October 19–20, 1948, Retlaw Hotel—Fond du Lac, Wisconsin (Madison, Wisc.: Cantwell Press, 1913/1915–1954), 28–29, accessed August 8, 2006 at www.digital.library.wisc. edu/1711.dl/WI.WCA.1948; Warren Resh, "Address," in *Proceedings of the Wisconsin Cheese Maker's Association Sixtieth Annual Meeting*, October 24–25, 1951, Retlaw Hotel and County Building, Fond du Lac, Wisconsin (Madison, Wisc.: Cantwell Press, 1913/1915–1954), 83–87, accessed August 8, 2006, at www.digital.library.wisc.edu/1711. dl/WI.WCA.1951.

140. Carmichael, "Forty Years of Water Pollution Control in Wisconsin," 418–419.

141. Robert Gottleib, *Forcing the Spring: The Transformation of the American Environmental Movement* (Washington, D.C.: Island Press, 1993), 55–59; Samuel Hays, *Beauty, Health, and Permanence: Environmental Politics in the United States, 1955–1985* (New York: Cambridge University Press, 1987), 22–26; Gregg Mitman, "In Search of Health: Landscape and Disease in American Environmental History," *Environmental History* 10 (April 2005): 184–211.

142. Carmichael, "Forty Years of Water Pollution Control in Wisconsin," 402–403; C. M. Stauffer to Alexander Wiley, February 15, 1957, folder Dairy Products—January 1–May 31, 1957, box 2911, RG 16, SOA, GC, 1906–70, NARA.

Chapter 4

1. Cele Roberts, interviewed by Brett Harvey, in *The Fifties: A Woman's Oral History* (New York: Harper Collins, 1993), 113–114.

2. Lorna Beall Smith (daughter of L. Dale Beall), in discussion with the author, December 24, 2007; Dale Beall to Quality Milk Association, Mount Carroll, Illinois, July 23, 1956, letter in 1954–1955 Illinois Farm Record Book in possession of the author.

3. Rosalyn Baxandall and Elizabeth Ewen, *Picture Windows: How the Suburbs Happened* (New York: Basic, 2000), 166–167.

4. Lizabeth Cohen, *Consumer's Republic: The Politics of Mass Consumption in Postwar America* (New York: Knopf, 2003); Kenneth Jackson, *Crabgrass Frontier: The Suburbanization of the United States* (New York: Oxford University Press, 1985); John Brinkerhoff Jackson, *Discovering the Vernacular Landscape* (New Haven: Yale University Press, 1986); Adam Rome, *Bulldozer in the Countryside: Suburban Sprawl and the Rise of American Environmentalism* (New York: Cambridge University Press, 2001); Dolores Hayden, *Building Suburbia: Green Fields and Urban Growth, 1820–2000* (New York: Pantheon, 2003); Alexander Wilson, *The Culture of Nature: North American Landscape from Disney to the Exxon Valdez* (Malden, Mass.: Blackwell, 1992).

5. Shane Hamilton's *Trucking Country: The Road to America's Wal-Mart Economy* offers the keenest analysis of the impact of the changing politics and economics of the postwar era on the nation's farms. This chapter is deeply indebted to the insights it provides. See Shane Hamilton, *Trucking Country* (Princeton: Princeton University Press, 2006), 99–101, 111–118.

6. Clyde Brunner, 1959 Annual Meeting Minutes, box 3, CBC Papers, WHS; Harry Jollie, Ice Cream Mix Manager, Statement at 1959 Annual Membership Meeting, Consolidated Badger Cooperative, box 3, CBC Papers, WHS.

7. *Ice Cream Field* (August 1950): 44; Carvel Ice Cream Records, 1934–1989, box 1, Accession 488, Archives Center, NMAH, Smithsonian Institution; "Dairy Queen Output Was Consumed at Pint Per Capita Rate in 1953," *Ice Cream Trade Journal* 50, no. 6 (June 1954): 110.

8. Hayden, *Building Suburbia*, 132.

9. I have found no quantitative data documenting the diversity of the ice cream–eating public, but the food's regular appearance in publications that cut across socioeconomic and racial lines and in locales that attracted diverse audiences suggest its widespread consumption. For instance, Good Humor peddled its wares on Coney Island and in Manhattan. Good Humor Ice Cream Collection, 1930–1980, Accession 451, box 1, folder 8, NARA, Smithsonian Institution; "Homemade Ice Cream," *Ebony* 8 (August 1953): 92–96; "Ice Cream," *Ebony* 9 (August 1954): 76–80; "Ice Cream Desserts," *Ebony* 15 (August 1960): 118, 120, 122; "Cherry Nugget Ice Cream," *Ebony* 16 (February 1961): 105; "Sealtest Ice Cream Bars," *Ebony* 17 (August 1962): 45; "Homemade Ice Cream," *Ebony* (August 1970): 174, 176, 178; civil rights boycotts focused on the food are documented in *Jet* on September 28, 1961, and August 19, 1965.

10. Fred Fisher, *Over a Chocolate Sundae on a Saturday Night* (New York: Fred Fisher Music Co., 1935); Pauline Arnold, *Won't You Have an Ice Cream Soda with Me?* (Burlington, Vt.: Pauline Arnold, 1954); John Redmond and Duke Leonard, *Tony Spumoni—the Ice Cream Man* (New York: Novelty Music Company, 1947). All in box 541B, Ice Cream, de Vincent Collection, Archives Center, NMAH; "The Soda Fountain Can—And Should—Be a Community Meeting Place," *American Druggist* 122, no. 1 (July 1950): 143–144.

11. Ronald Kline, *Consumers in the Country: Technology and Social Change in Rural America* (Baltimore: Johns Hopkins University Press, 2000), 201.

12. See Hendler's Ice Cream folder, in Jules Frandsen and D. Horace Nelson, *Ice Creams and Other Frozen Desserts* (Amherst, Mass.: J. H. Frandsen, 1950), 164–165; Creamery Package Manufacturing Company, *The Handy Book of Dairy Supplies, Catalog No. 42* (Chicago: Creamery Package Manufacturing Company, 1942), 139, Hagley Museum and Library,

Trade Catalog Collection; Grier Lowry, "Drug Store with a 'Party' Line," *American Druggist* 122, no. 4 (October 1950): 174.

13. Boyd Sherwood, Sunday, July 4, 1943, box 2, item 14, Boyd Sherwood Diaries, 1927–1955, SC22581, New York State Library and Manuscripts Collection, Albany, N.Y.

14. Donald McCormick, "The Maine Family Reunion," November 26, 1941, box A831, U.S. Work Projects Administration Records, America Eats Papers, Library of Congress, Washington, D.C.

15. Ice Cream Trade Journal, "The Trend Is Favorable for Greater Ice Cream Consumption," *Ice Cream Trade Journal* 50 (October 1954): 112; between 1956 and 2009, per capita consumption of total frozen dairy products (including low-fat ice cream and frozen yogurt) rivaled or exceeded the 1946 level, but 1946 remains the high point for ice cream consumption, specifically. Economic Research Service (ERS), U.S. Department of Agriculture, Food Availability (Per Capita) Data System, accessed April 7, 2010, available at http://www.ers.usda.gov/Data/FoodConsumption.

16. Elaine Tyler May, *Homeward Bound: American Families in the Cold War Era: American Families in the Cold War Era*, rev. ed. (New York: Basic, 1999), 147; Arnold Blumenthal, "Latest Study of the Nation's Home Storage Capacity for Ice Cream," *Ice Cream Trade Journal* 48 (December 1952): 26.

17. Harold King, "There's No Place Like Home," *Ice Cream Field* 56, no. 2 (August 1950): 38, 60.

18. Ice Cream Trade Journal, "Analysis of the Industry's Volume in Packages, Cups, and Novelties," *Ice Cream Trade Journal* 52, no. 8 (August 1956): 13.

19. Editor, "The Role of the Soda Fountain," *Ice Cream Trade Journal* 48, no. 4 (April 1952): 20.

20. "Kelvinator Promotes Ice Cream Sales," *Ice Cream Field* 55 (June 1950): 105.

21. Radio advertisements, 1962, Eskimo Pie Collection, 1921–1992, box 25, Archives Center, NMAH.

22. Carvel Ice Cream Records, 1934–1989, box 8, Archives Center, NMAH, SI.

23. Marian Moore and Julia Pond, "Dessert Choices of Southern Urban Families," *Journal of Home Economics* 58 (October 1966): 660; May, *Homeward Bound*, 149.

24. "Supermarket Potential," *Ice Cream Trade Journal* 48 (March 1952): 36.

25. M. M. Zimmerman, "Mass Selling and Shopping," *Ice Cream Field* 56 (September 1950): 22.

26. E. I. DuPont de Nemours and Company, *Design for Selling: A Study of Impulse Buying* (Wilmington, Del.: E. I. DuPont de Nemours, 1946), 3.

27. F. A. Babione, "Will She Buy Your Brand?" *Ice Cream Field* 63 (April 1954): 58–60.

28. H. L. Darnsaedt to W. A. Greiner, June 20, 1955, box 4; Eskimo Pie Display Cabinet, Box 10; Eskimo Pie Company Annual Reports, 1954, 1955, 1957, box 11; all in Eskimo Pie Records, 1920–1992, Archives Center, NMAH.

29. B. Alene Theisner, "What Customers Say," *Ice Cream Field* 63 (1954): 50.

30. Frank Elliott, Washington, D.C., to FDA, August 9, 1956, FDA General Subject Files, 1938–1974, folio 2128, RG 88, NARA II, College Park.

31. Lawrence Little, "Stabilizers and Emulsifiers," *Ice Cream Field* 55 (February 1950): 68–71; "Shervel" ice cream stabilizer debuted in 1949. Creamery Package Manufacturing Company, *Catalog of Supplies and Equipment for Dairy and Food Processing Industries: Catalog 1903* (Chicago: Creamery Package Manufacturing Co., 1949), 93; "Looking Back on a Quarter of a Century of Advancements in Ice Cream Research," *Ice Cream Trade Journal* 48 (April 1952): 34, 111.

32. Vincent Rabuffo, "Vegetable Fat Frozen Desserts?" *Ice Cream Trade Journal* 48 (November 1952): 20–22, 24, 94–97; "By the Editor," *Ice Cream Trade Journal* 48 (June 1952): 18, 72.

33. National Association of Ice Cream Manufacturers, "Comparative State Standards for Ice Cream, Frozen Custard, Sherbets, and Ices," 1958 in FDA General Subject Files, 1938–1974, box 2487, RG 88, NARA II, College Park, Md.

34. Hamilton, *Trucking Country*, 128–134.

35. FDA press release, March 26, 1958, in FDA General Subject Files, 1938–1974, box 2487, RG 88, NARA II, College Park, Md.

36. FDA press release, March 26, 1958, in FDA General Subject Files, 1938–1974, box 2487, RG 88, NARA II, College Park, Md.; for more on the history of FDA standards of identity, see Alissa Hamilton, *Squeezed: What You Don't Know about Orange Juice* (New Haven: Yale University Press, 2009), 31–61.

37. Frank Elliott to FDA, August 9, 1956, in FDA General Subject Files, box 2128, RG 88, NARA II, College Park, Md.

38. Susan Keegan to Jim Delaney, March 5, 1955, in FDA General Subject Files, 1938–1974, box 1959, RG 88, NARA II, College Park, Md.

39. A. S. McCully to FDA, March 19, 1956; Naomi Comyn to FDA, May 7, 1957; Mrs. P. M. Horwitz to FDA, June 16, 1958; all in FDA General Subject Files, 1938–1974, box 2307, RG 88, NARA II, College Park, Md.

40. Ralph Harford to Charles Potter, U.S. Senator, March 24, 1954; Roy Garrison to FDA, June 23, 1955; Hugh Hartman to Ezra Taft Benson, February 2, 1955; William Jackson to FDA, July 28, 1956; Mrs. Ervin Bell to Department of Agriculture, July 13, 1956; Mrs. Edna Nolan, May 22, 1957. All in FDA General Subject Files, 1938–1974, box 1959, RG 88, NARA II, College Park, Md.

41. Theodora Burke to FDA, October 20, 1958, box 2487, RG 88, FDA General Subject Files, 1938–1974, NARA II, College Park, Md.

42. Stella Durrell to FDA, July 28, 1958; Ursula Tuohy to Senator Prescott Bush, April 21, 1958; FDA to Mr. Robert Martin, August 25, 1958, responding to enclosure from "Nature's Path magazine," box 2487, RG 88 FDA General Subject Files, 1938–1974.

43. On desserts and gender in the 1950s, see Karal Ann Marling, *As Seen on TV: The Visual Culture of Everyday Life in the 1950s* (Cambridge: Harvard University Press, 1994), 226–227; Harvey Levenstein, *Paradox of Plenty: A Social History of Eating in Modern America*, rev. ed. (Berkeley: University of California Press, 2003), 124–125; Katherine Parkin, *Food Is Love: Food Advertising and Gender Roles in Modern America* (Philadelphia: University of Pennsylvania Press, 2006), 36–37; Anne Vilesis, *Kitchen Literacy: How We Lost Knowledge of Where Food Comes From and Why We Need to Get It Back* (Washington, D.C.: Island Press, 2008), 191–193.

44. Jacobs, *Purchasing Power*, 124–125.

45. Donald Davidson, "How Manufacturing Co-ops Market Grade A Milk," Circular 26 (Washington, D.C.: U.S. Department of Agriculture, Farm Cooperative Service, 1960), 4.

46. Wawa Dairy Farms, 1903–1958, Millville Manufacturing Company Records, Series 3, Sub-series A, box 6, Accession 1772, Hagley Museum and Library, Wilmington, Del.

47. "Five-State Reciprocity for Milk Inspection Roundtable," *Twenty-Second Annual Report of the New York State Association of Milk Sanitarians, 1948* (Albany: New York State Association of Dairy Inspectors, 1949), 51–64.

48. Table 21—Farms Reporting Milk Cows, by Number on Hand, By Divisions and States, 1954 and 1950, U.S. Bureau of the Census, *United States Census of Agriculture, 1954*, vol. 2, General Report, Statistics by Subjects (Washington, D.C.: Government Printing Office, 1956), 476–477.

49. R. N. Van Arsdall, D. B. Ibach, and Thayer Cleaver, *Economic and Functional Characteristics of Farm Dairy Buildings. Bulletin 570* (Urbana: University of Illinois Press, 1953), 3, 13.

50. Ray Hoglund and Joe Marks, "Suburbs Crowding Your Farm?" *Hoard's Dairyman* 109, no. 1 (January 10, 1964): 11, 39; Mr. and Mrs. L. J. Totten to Vice President Richard Nixon, June 4, 1958, box 3098; Leo Miner to Ezra Taft Benson, April 21, 1955, box 2558, Folder Dairy Products May 5–June 10, 1955; both in RG 16, SOA, GC, 1906–76, NARA; John

Fraser Hart, *The Look of the Land* (Englewood Cliffs, N.J.: Prentice Hall, 1975), 120–121.

51. L. C. Cunningham and R. C. Wells, "Changes in Commercial Dairy Farming, Central Plain Region, New York 1954–1964," *Cornell University Agricultural Experiment Station Bulletin 1013* (Ithaca: Cornell University and New York State Department of Agriculture, 1967), 4–5.

52. USDA figures, cited in Steven Hoffbeck, *The Haymakers: A Chronicle of Five Families* (St. Paul: University of Minnesota Press, 2000), 146.

53. D. N. McDowell, "The Milk House Requirement," *Milk Dealer* 42 (November 1952): 104–107.

54. "Bulk Milk Handling Panel," 1956 Annual Meeting of Consolidated Badger Cooperative, CBC Papers, Mss 104, box 3, folder 10; United States Steel Corporation, *Bulk Handling of Milk with Stainless Steel* (Pittsburgh: United States Steel Corporation, 1955): 2–24; Hagley Museum and Library, Trade Catalog Collection, Wilmington, Del.; Noel Stocker, "Progress in Farm-to-Plant Bulk Milk Handling," *Farm Cooperative Service Circular 8* (Washington, D.C.: U.S. Department of Agriculture, November 1954), 4–5.

55. Stocker, "Progress in Farm-to-Plant Bulk Milk Handling," 4–5, 36–37.

56. Herbert Leon to Senator Clyde Wheeler, April 2, 1956, box 2793, RG 16, SOA, GC, 1906–76, NARA.

57. "We Like Bulk Handling," *Hoard's Dairyman* 99 (March 10, 1954): 224–225.

58. Lu Ann Jones interview of Ralph, Dorothy and Bobby Lewis, May 2, 1987, transcript, pp. 72–73, Southern Agriculture Oral History Project, Archives Center, National Museum of American History, Smithsonian Institution, Washington, D.C.

59. S. A. Witzel and E. E. Heizer, "Loose Housing or Stanchion Type Barns? Summary of a 10 Year Dairy Cattle Housing Experiment in Southern Wisconsin," *Bulletin 503* (Madison: University of Wisconsin Experiment Station, 1953), 3, 17; H. J. Barre and L. L. Sammet, *Farm Structures* (New York: John Wiley, 1950), 225–227; A. M. Goodman, "Remodeling Barns for Better Dairy Stables," *Cornell Extension Bulletin 742* (Ithaca, November 1951): 7.

60. Witzel and Heizer, "Loose Housing or Stanchion Type Barns?," 17.

61. Barre and Sammet, *Farm Structures*, 226–227; Nils-Ivar Isaksson, "Economies of Scale in Milk Production and the Competitive Position of the Family Farm" (Master's thesis, University of Wisconsin, 1961): 78. Despite experts' approval, farmers' adoption of loose housing was slow. A 1963–1964 survey of dairy farms in central New York found that 90 percent of farms still housed milk cows in stall barns; only among the larger farms did loose housing make up half of the barn arrangements. Frank Hill, interview with Lu Ann Jones, May 4, 1987, Southern Agriculture Oral History Project, box 18, Accession 773, Archives Center, NMAH; Cunningham and Wells, "Changes in Commercial Dairy Farming," 10–11.

62. Cunningham and Wells, "Changes in Commercial Dairy Farming," 7; see also U.S. Dairy Barn Project RMe550, September 11, 1952, attached to D. Howard Doane to O. E. Reed, box 12, Record Group 152, Records of the Bureau of Dairy Industry, National Archives.

63. Loren Neubauer and Harry Walker, *Farm Building Design* (Englewood Cliffs, N.J.: Prentice Hall, 1961), 57–61; Barre and Sammet, *Farm Structures*, 217–225; Dairy Cattle Housing Subcommittee of the North Central Regional Farm Buildings Committee, "Dairy Cattle Housing in the North Central States," *Bulletin 470* (Madison, Wisc.: Agricultural Experiment Station, 1947), 47–50.

64. Stocker, "Progress in Farm-to-Plant Bulk Milk Handling," 44; Randolph Barker and Earl Orel Heady, *Economy of Innovations in Dairy Farming and Adjustments to Increase Resource Returns* (Ames: Iowa State University, 1960), 747–764.

65. Max Foster to Ernest Greenough, February 1963, folder 6.13 Testing, box 6, Sunshine Dairy Farm Records, D-385, University of California-Davis Special Collections, Davis, California.

66. Hank Bennett to Ernest Greenough, February 14, 1963, and article clipping, Web Allison, "Automating Bossy," both in box 6, Sunshine Dairy Farm Records, D-385, University of California Davis Special Collections, Davis, California.

67. Elihu Gifford Papers, May 13, 1910, May 6, 1911, May 24, 1917, May 13, 1919, May 20, 1920, May 14, 1923, May 17, 1924, May 26, 1925; items 4–7, box 1, SC 18734, New York State Manuscripts and Special Collections, Albany, N.Y.

68. Michael Pollan, *The Omnivore's Dilemma: A Natural History of Four Meals* (New York: Penguin, 2006), 65–84.

69. C. R. Hoglund, "High Quality Roughage Reduces Dairy Costs," *Michigan State Agricultural Experiment Station Bulletin 390* (East Lansing: Michigan State College, February 1954), 3–4; Karl Vary, "Economics of Grassland Farming," *Michigan State Agricultural Experiment Station Bulletin 391* (East Lansing: Michigan State College, May 1954), 5–10.

70. "I Prefer to Graze My Herd," *Hoard's Dairyman* 101, no. 14 (July 25, 1956): 721.

71. Lynne Heasley, *A Thousand Pieces of Paradise: Landscape and Property in the Kickapoo Valley* (Madison: University of Wisconsin Press, 2005), 52–55.

72. Ford S. Prince, "A Revolution in Dairying," *Hoard's Dairyman* (February 25, 1952): 174–175.

73. "Our Cows Like Zero Grazing," 176–177; C. R. Hoglund, "I Haul Pasture to My Cows," *Hoard's Dairyman* (1956): 720.

74. Harold Owen in "Our Cows Like Zero Grazing," *Hoard's Dairyman* 103, no. 4 (February 25, 1958): 177; Foster to Greenough, February 1963, box 6, Sunshine Farm Dairy Records, MS D-385, University of California-Davis, Davis, California.

75. Hoffbeck, *The Haymakers*, 149–150, 153–54; Susan Granger and Scott Kelly, *Historic Context Study of Minnesota Farms, 1820–1960*, vol. 1 (Morris: Minnesota Department of Transportation, June 2005), 3.113; Kenneth L. Smith (former Harvestore salesman) in conversation with the author, September 1, 2009.

76. "Our Roughage Feeding Program," *Hoard's Dairyman* 97, no. 5 (March 10, 1952): 220–221, 230–233.

77. N. N. Allen, "Questions from the Mailbox," *Hoard's Dairyman* 103, no. 7 (April 10, 1958): 379.

78. Allen, "Questions from the Mailbox," 379.

79. For the ideas of nature that brought suburbanites to the countryside, see Walter McKain Jr., "The Exurbanite: Why He Moved," in *A Place to Live: The Yearbook of Agriculture, 1963* (Washington, D.C.: U.S. Department of Agriculture, 1963), 26–29.

80. Daniel Horowitz, *The Anxieties of Affluence: Critiques of American Consumer Culture, 1939–1979* (Amherst: University of Massachusetts Press, 2004), 101–150.

81. See, for instance, Letters to the Editor, *Hoard's Dairyman* 101, no. 8 (April 25, 1956): 403–404.

82. J. Van Dam and C. Perry, "Manure Management: Costs and Product Forms," *California Agriculture* (December 1968): 12–13.

83. Glenn Voskuil, *Liquid Manure Systems for the Dairy* (Merced: University of California Extension Service, 1962), 2; Hank Bennett to Ernest Greenough, February 14, 1963, box 6, D-385, University of California-Davis Special Collections, Davis, California.

84. J. Ronald Miner, "Farm Animal-Waste Management," North Central Regional Publication 206, Special Report 67. Agricultural Experiment Stations of AL, IL, IN, IA, KS, MI, MN, MO, NB, ND, OH, SD, WI, and USDA (Ames: Iowa State Experiment Station, May 1971), 25–31; Samuel A. Hart, "Manure Management," *California Agriculture* (December 1964): 5–7.

85. Greenough to Mr. Henry Bennett, January 19, 1965; Bennett to Greenough, January 20, 1965, February 8, 1965; Sunshine Farm Dairy Records, box 6, University of California-Davis, Davis, California.

86. Miner, "Farm Animal-Waste Management," 33.

87. Miner, "Farm Animal-Waste Management," 33; Anderson, *Industrializing the Corn Belt*, 131–139.

88. Hart, "Manure Management," 7.

89. Kurt Greenlund et al., "Heart Disease and Stroke Mortality in the Twentieth Century," in *Silent Victories: The History and Practice of Public Health in Twentieth-Century America*, ed. John Ward and Christian Warren, 381 (New York: Oxford University Press, 2007).

90. "Transcript of Heart Specialist's News Conference on President's Health," *New York Times*, September 27, 1955, p. 21.

91. Robert K. Plumb, "Heart Affliction Related to Stress," *New York Times*, March 9, 1958. Walter Adams, "The 100% American Way to Die," *Better Homes and Gardens* 26 (November 1947): 224–227; Robert K. Plumb, "Stress of Work in United States Culture Cited as Cause of Heart Disease," *New York Times*, January 28, 1959; Barbara Ehrenreich, *The Hearts of Men: American Dreams and Flight from Commitment* (New York: Doubleday, 1983), 70–71; Harold Schmeck, "'Rags to Riches' Hard on Health," *New York Times*, May 2, 1960; Paul Dudley White, "Rx for Health: Exercise," *New York Times*, June 23, 1957, p. 189; "Hard Exercise Urged: 2 Experts Say Schools Must Press Fitness Program," *New York Times*, September 29, 1961, p. 35; Oglesby Paul, *Take Heart: The Life and Prescription for Living of Dr. Paul Dudley White* (Cambridge: Harvard University Press, 1986), 192.

92. Ancel Keys, "Human Atherosclerosis and the Diet," *Circulation* 5 (1952): 115–118.

93. R. K. P., "Diet Related to Heart Disease," *New York Times*, September 19, 1954.

94. Philip Benjamin, "Steep Drop in Milk Consumption Stirs Government and Dairymen," *New York Times*, February 18, 1962; "Panel Doubts Fat Spurs Heart Ills," *New York Times*, November 21, 1956; Warren Weaver, "2 Doubt Fat's Tie to Heart Disease," *New York Times*, July 16, 1957; "Heart Association Experts Doubt That Fats Diet Hardens Arteries," *New York Times*, August 15, 1957; William Laurence, "Science in Review: More Light on Supposed Relation of Fats in the Diet to Prevalent Heart Disease," *New York Times*, January 12, 1958.

95. "Heart Unit Backs Reduction in Fat," *New York Times*, December 11, 1960.

96. "Special Report: Heart Attacks: Can Diet Prevent Them?" *Good Housekeeping* 151 (May 1961): 138; Karin Garrety, "Social Worlds, Actor-Networks, and Controversy: The Case of Cholesterol, Dietary Fat, and Heart Disease," *Social Studies of Science* 27, no. 5 (October 1997): 740–741; "Dairy Leaders to Diet, with an Accent on Milk," *New York Times*, January 30, 1963.

97. American Dairy Association Cholesterol Committee Report, July 21, 1953, box 2, Records of the Bureau of Dairy Industry, Correspondence and Other Records, 1944–1953, Record Group 152, NARA.

98. Garrety, "Social Worlds, Actor Networks, and Controversy," 741; Peter Bart, "Advertising: Dairy Men Open Counter Attack," *New York Times*, August 7, 1962.

99. "Dairy Leaders to Diet, with an Accent on Milk," 15.

100. W. C. Callender, "To Sell to More Homes," *Ice Cream Field* 63 (February 1954): 46; Frandsen and Nelson, *Ice Cream and Other Frozen Desserts*, 8.

101. "Dietetic Frozen Desserts," *Ice Cream Trade Journal* 52 (March 1956): 14–15, 136–139.

102. "Latest Analysis of Soft-Serve and Counter Freezer Gallonage," *Ice Cream Trade Journal* (March 1956): 39.

103. Jack Gould, "TV: C.B.S. Studies 'The Fat American,'" *New York Times*, January 19, 1962.

104. George Barnes to Secretary Freeman, January 22, 1962, box 3748, RG 16, SOA, GC, 1906–76, NARA; for a contemporary version of this problem, see Tara Parker-Pope, "Why the Government Is Promoting Cheese," *New York Times*, November 7, 2010.

105. Garrety, "Social Worlds, Actor-Networks, and Controversy," 742.

106. George Rupple, 1962 Annual Meeting, Consolidated Badger Cooperative, 17, box 3, CBC Papers, WHS.

107. Paul, *Take Heart*, 198–199.

108. Good Humor Ice Cream Collection, box 1, Archives Center, NMAH.

Chapter 5

1. Mrs. Frederick Faust and others to Dr. J. Earl Smith, October 26, 1956, box 1, Committee for Environmental Information Records, SL 69, Western Historical Manuscript Collection, University of Missouri, Saint Louis (hereafter CEI Records, SL 69, WHMC, UMSL); "Seek to Determine If 'Fall-Out' Affects Milk," *St. Louis Globe Democrat*, October 30, 1956.

2. W. F. Libby, "Radioactive Fallout and Radioactive Strontium," *Science* 123 (April 20, 1956): 658; Harold Knapp, "The Effects of Deposition Rate and Cumulative Soil Level on the Concentration of Strontium-90 in U.S. Milk," Division of Biology and Medicine, U.S. Atomic Energy Commission, Division of Technical Information (Washington, D.C.: Government Printing Office, 1961), 1.

3. Gertrude Faust to Chet Holifield, chairman on the special subcommittee on radiation, May 29, 1957, CEI Records, SL 69, WMHC, UMSL.

4. Robert Divine, *Blowing on the Wind: The Nuclear Test Ban Debate, 1954–1960* (New York: Oxford University Press, 1978), 104–105; for footage of Johnson advertisement, see Museum of the Moving Image, "The Living Room Candidate: Presidential Campaign Commercials, 1952–2008," Museum of the Moving Image, available at http://www.livingroomcandidate.org/commercials/1964, accessed October 6, 2008.

5. "The Milk All of Us Drink—and Fallout," *Consumer Reports* 24 (March 1959): 102–103.

6. C. S. Bryan, "Prevention and Control of Mastitis," *Hoard's Dairyman* 86 (March 10, 1941): 146, 177; "Herd Health: Mastitis," *Hoard's Dairyman* 92 (February 25, 1947): 135, 163; Ival Merchant and R. Allen Packer, *Handbook for the Etiology, Diagnosis, and Control of Infectious Bovine Mastitis*, 2nd ed. (Minneapolis: Burgess Publishing, 1952), 2; H. E. Calbert, "The Problems of Antibiotics in Milk," *Journal of Milk and Food Technology* 14 (March–April 1951): 63–64.

7. Gertrude Mallary to Mr. A. L. Hamelin, February 12, 1946, Mallary Farm Records, VT HS; Edward Meigs et al., "The Relationship of Machine Milking to the Incidence and Severity of Mastitis," *U.S. Department of Agriculture, Technical Bulletin Number 992* (Washington, D.C.: Government Printing Office, 1949), 1–51; Daniel Noorlander and David Gray, "Section II: Milking Machines and Mastitis, Third Edition," in *Mechanics and Production of Quality Milk*, ed. Daniel Noorlander, David Gray, and John Dahl, 40, 54–55 (Madison, Wisc.: Democrat Printing Company, 1965).

8. Terry Summons, "Animal Feed Additives, 1940–1966," *Agricultural History* (1968): 305–313; William Boyd, "Making Meat: Science, Technology, and American Poultry Production," *Technology and Culture* 42 (October 2001): 646–649; Mark Finlay, "Hogs, Antibiotics, and the Industrial Environments of Postwar Agriculture," in *Industrializing Organisms: Introducing Evolutionary History*, ed. Philip Scranton and Susan Schrepfer, 242–246 (New York: Routledge, 2004).

9. American Cyanamid Company advertisement for Targot Mastitis Ointment, *Dairymen's League News* 40 (January 3, 1956): 20; Lederle Laboratories Aereomycin advertisement, *American Druggist* (April 14, 1952): 64; "Many Brands of Mastitis Preps Displayed to Meet All Demands," *American Druggist* 137 (June 2, 1958): 68.

10. S. F. Scheidy, "Antibiotic Therapy in Veterinary Medicine," *American Veterinary Medical Association Journal* 118 (April 1951): 213–220; H. D. Kautz, "Penicillin and Other Antibiotics in Milk," *Journal of the American Medical Association* 171 (September 5, 1959): 135–137; J. L. Albright, S. L. Tuckey, and G. T. Woods, "Antibiotics in Milk: A Review," *Journal of Dairy Science* 44 (May 1961): 787–788; Henry Welch, William Jester, and J. M. Burton, "Antibiotics in Fluid Market Milk: Third Nationwide Study," *Antibiotics and Chemotherapy* 6 (May 1956): 369–374; Henry Welch, "Control of Antibiotics in Food," *Food, Drug, and Cosmetics Journal* 12 (1957): 464.

11. Ralph Little and Wayne Plastridge, eds., *Bovine Mastitis: A Symposium* (New York: McGraw-Hill, 1946), 413; American Veterinary Medical Association, "Penicillin Has Limitation," *Hoard's Dairyman* 92, no. 14 (July 10, 1947): 585; Daniel Noorlander et al., *Mechanics and*

Production of Quality Milk (Madison, Wisc.: Democrat Printing, 1965), 280; American Foundation for Animal Health advertisement, *Hoard's Dairyman* 92, no. 13 (June 25, 1947): 536.

12. Alec Bradfield, L. A. Resi, and D. B. Johnstone, "Presence of Aureomycin in Milk and Its Effect on Cheese Production," *Journal of Dairy Science* 35 (January 1952): 51, 56–57; H. Katznelson and E. G. Hood, "Influence of Penicillin and Other Antibiotics on Lactic Streptococci in Starter Cultures Used in Cheddar Cheesemaking," *Journal of Dairy Science* 32 (November 1949): 961–968; C. S. Bryan, "Problems Created for the Dairy Industry by Antibiotic Mastitis Treatments," *Journal of Milk and Food Technology* 14 (September–October 1951): 161–162; Robert Hardell to O. E. Reed, Bureau of Dairy Industry, September 1, 1953, box 5, folder Cheese 3, Record Group 152, Records of the Bureau of Dairy Industry, 1944–1953, National Archives and Records Administration, College Park, Md. (hereafter BDI, RG 152, NARA).

13. For shifts to bulk handling, see Joseph Cowden, "Comparing Bulk and Can Milk Hauling Costs," *Farm Cooperative Service Circular 14* (Washington, D.C.: U.S. Department of Agriculture, 1956); United States Steel Corporation, *Bulk Handling of Milk with Stainless Steel* (Pittsburgh: United States Steel Corporation, 1955), 2–24, Hagley Museum and Library, Trade Catalog Collection, Wilmington, Del.; G. A. Claybaugh and F. E. Nelson, "The Effect of Antibiotics in Milk: A Review," *Journal of Milk and Food Technology* 14 (September–October 1951): 157–158; H. E. Calbert, "The Problem of Antibiotics in Milk," *Journal of Milk and Food Technology* 14 (March–April 1951): 63.

14. Wisconsin Cheese Makers' Association, "Resolution Number Seven," in *Proceedings of the Wisconsin Cheese Makers' Association Sixtieth Annual Meeting*, October 24–25 (Fond du Lac: Wisconsin Cheese Makers' Association, 1951), 43–44.

15. Administrative Procedure Act, Title 21, § 3.25, Federal Register, 16 (August 21, 1951): 8270.

16. Reprinted in memo by A. M. Meekman, Dairy Husbandman, USDA Federal Extension Service to all County Agricultural Agents, December 28, 1959, box 2818, FDA General Subject Files, RG 88, NARA.

17. Welch, "Control of Antibiotics in Food," 466–467; H. R. Vickers, L. Bagratuni, and S. Alexander, "Dermatitis Caused by Penicillin in Milk," *Lancet* 1 (1958): 377; H. D. Kautz, "Penicillin and Other Antibiotics in Milk," *JAMA* 171 (September 5, 1959): 135–137. An Oxford unit is equivalent to .6 micrograms. M. R. Stephens, "Policies and Programs Concerning Antibiotic and Pesticide Residues in Milk," *Journal of Dairy Science* 43 (April 1960): 581.

18. Ray Clark, Assistant Manager, Carnation Company to Mr. George Clarke, Dairy Products Institute, February 16, 1960, box 2818; Kraft Foods pamphlet, box 2821; Clyde Russell, Group Discussion Held at Michigan State University, December 22, 1959, and Nevis Cook, Boston District FDA Report, December 11, 1959, box 2650; all in FDA General Subject Files.

19. Cliff Shane to Kansas City District, January 7, 1959, box 2650; and Jerome Trichter to George Larrick, March 17, 1970, box 2821, FDA General Subject Files.

20. Wesley Sawyer to Oren Harris, March 22, 1960; and K. M. Autrey to George P. Larrick, May 31, 1960, box 2821, FDA General Subject Files.

21. Dr. Lyle Alexander, USDA, in *Hearings*, "The Nature of Radioactive Fallout and Its Effects on Man," Joint Committee on Atomic Energy, Eighty-Fourth Congress, First Session (Washington, D.C.: Government Printing Office, 1957), 514; for SR-90 generally, see Paul S. Boyer, *Fallout: A Historian Reflects on America's Half-Century Encounter with Nuclear Weapons* (Columbus: Ohio State University, 1998), 82, 85; J. Samuel Walker, *Permissible Dose: A History of Radiation Protection in the Twentieth Century* (Berkeley: University of California Press, 2000), 21–22; Carolyn Kopp, "The Origins of the American Scientific Debate over Fallout Hazards," *Social Studies of Science* 9 (November 1979): 413–415;

Michael Egan, *Barry Commoner and the Science of Survival: The Remaking of American Environmentalism* (Cambridge: MIT Press, 2007), 52.

22. Samuel P. Hays, *Beauty, Health, and Permanence: Environmental Politics in the United States, 1955–1985* (New York: Cambridge University Press, 1987), 24–25.

23. Egan, *Barry Commoner and the Science of Survival*, 60–61; Ralph Lutts, "Chemical Fallout: Rachel Carson's *Silent Spring*, Radioactive Fallout, and the Environmental Movement," *Environmental Review* 9 (Fall 1985): 212–225; Adam Rome, "'Give Earth a Chance': The Environmental Movement and the Sixties," *Journal of American History* 90 (September 2003): 534–539; Adam Rome, *The Bulldozer in the Countryside: Suburban Sprawl and the Rise of Environmentalism* (New York: Cambridge University Press, 2001), 7.

24. Lawrence Glickman, *Buying Power: A History of Consumer Activism in America* (Chicago: University of Chicago Press, 2009), 266.

25. History of the Saint Louis Consumer Federation, box 1, folder 12, St. Louis Consumer Federation Papers, 1935–1977, SL 395, Western Historical Manuscripts Collection, University of Missouri, St. Louis; Meg Jacobs, *Pocketbook Politics: Economic Citizenship in Twentieth-Century America* (Princeton: Princeton University Press, 2005), 122–127; Lizbeth Cohen, *A Consumers' Republic: The Politics of Mass Consumption in Postwar America* (New York: Knopf, 2003), 28–35.

26. Statement of Mrs. Frederick Faust, December 12, 1956, CEI Records, SL 69, box 1, WHMC, UMSL; William Cuyler Sullivan Jr., *Nuclear Democracy: A History of the Greater Saint Louis Committee for Environmental Information, 1957–1967* (St. Louis: Washington University Press, 1982), 15.

27. Edna [Mrs. George] Gellhorn, statement before the Subcommittee on Disarmament of the Senate Foreign Relations Committee, December 12, 1956, box 1, CEI Records, SL 69, WHMC, UMSL; Cohen, *Consumer's Republic*, 24.

28. Divine, *Blowing on the Wind*, 165–169.

29. Amy Swerdlow, *Women Strike for Peace: Traditional Motherhood and Radical Politics in the 1960s* (Chicago: University of Chicago Press, 1993), 83.

30. Steven Spencer, "Fallout: The Silent Killer, How Soon Is Too Late?" *Saturday Evening Post* 232 (September 29, 1959): 84; letter to the editor from Worried Mother, *New York Times*, March 23, 1959; "Serious Fallout Cases Found in the Midwest," *Washington Post*, June 7, 1959, pp. A1, A12–13.

31. Elaine Tyler May, *Homeward Bound: American Families in the Cold War Era* (New York: Basic, 1988), xxv–xxvi, 19–20.

32. Orville Freeman to Ronald McCamus, April 11, 1962, box 3748, folder Dairy Products, March 1–April 20, RG 16, SOA, GC, 1906–76, NARA; Harold Knapp, *The Effect of the Deposition Rate and Cumulative Soil Level on the Concentration of Strontium-90* (Washington, D.C.: U.S. Atomic Energy Commission, Division of Technical Information, 1961), 8, 33–35; Surgeon General to Mr. Curtis, House of Representatives, May 23, 1958, box 1, CEI Records, SL 69, WHMC, UMSL.

33. Sullivan, *Nuclear Democracy*, 1–20; Egan, *Barry Commoner and the Science of Survival*, 60–61; Lutts, "Chemical Fallout," 215–216.

34. John Fowler, ed., *Fallout: A Study of Superbombs, Strontium-90, and Survival* (New York: Basic, 1960), 53–55; Sullivan, *Nuclear Democracy*, 42–44; Egan, *Barry Commoner and the Science of Survival*, 65–68; lists of cities from press release in box 30, Scrapbooks, CEI Records, SL 69, WHMC, UMSL.

35. "The Milk All of Us Drink and Fallout," 108–111.

36. "Letters to the Editor," *Consumer Reports* 24 (May 1959): 219–220; "Letters to the Editor," *Consumer Reports* 24 (July 1959): 388.

37. Walter Eckelmann, J. Laurence Kulp, and Arthur Schulert, "Strontium-90 in Man II," *Science* 127 (February 7, 1958): 266–274.

38. Lester Machta, *Hearings*, "The Nature of Radioactive Fallout and its Effects on Man," 156–157; Divine, *Blowing on the Wind*, 131–132, 184–185.

39. Bruce Snow to Miles Horst, Assistant to Secretary of Agriculture, December 3, 1956, RG 16, box 2739, folder October 5, RG 16, SOA, GC, 1906–76, NARA.

40. Congresswoman Katherine St. George to Ezra Taft Benson, Secretary of Agriculture, January 30, 1957, RG 16, box 2910, folder June 1–August 1, 1957, RG 16, SOA, GC, 1906–76, NARA.

41. B. L. Larson and K. E. Ebner, "Significance of Strontium-90 in Milk: A Review," *Journal of Dairy Science* 41 (December 1959): 1651–1656.

42. "Scientists Urge Milk Be Kept in Diet in Spite of Strontium-90," *St. Louis Post Dispatch*, July 5, 1959, in box 30, Scrapbooks, CEI Records, SL 69, WHMC, UMSL; "The Milk All of Us Drink and Fallout," 111; response to letter to the editor, *Consumer Reports* 24 (July 1959): 388; "Fallout in Our Milk: A Follow-up Report," 65.

43. Greater St. Louis Citizens' Committee for Nuclear Information, May 10, 1962, box 1, Folder 11 Press Releases, CEI Records, SL 69, WHMC, UMSL.

44. U.S. Department of Health, Education, and Welfare, "Full-Scale System for Removal of Radiostrontium from Fluid Milk," Proceedings of Seminar, February 24–25, 1965, *Public Health Service Publication 999-RH-28* (Washington, D.C.: Government Printing Office, 1967), 8–13; "Test Clears Milk of Strontium-90," *New York Times*, December 20, 1961, 26; "Getting the Strontium-90 Out of the Milk," *Consumer Reports* (February 1962): 63.

45. Mrs. Phyllis (Kenneth L.) Smith to Orville Freeman, Secretary of Agriculture and Audrey (Mrs. William E.) Elder to Senator Estes Kefauver, February 5, 1962, box 3748, RG 16, SOA, GC, 1906–76, NARA; Edward Walsh to James J. Delaney, May 30, 1962, James J. Delaney Papers, Series 1, box 1, Special Archives and Collections, University at Albany, Albany, N.Y.

46. Andrew Kirk, *Counterculture Green: The Whole Earth Catalog and American Environmentalism* (Lawrence: University of Kansas Press, 2007), 6, 8–9; Ted Steinberg, "Can Capitalism Save the Planet? On the Origins of Green Liberalism," *Radical History Review* 107 (Spring 2010): 9–10.

47. "Method Taking Strontium-90 Out of Milk is Being Developed," *St. Louis Post-Dispatch*, June 30, 1959; press release, November 23, 1959, Box 1; "Research to Remove Strontium from Milk Started by Pevely," *St. Louis Post-Dispatch*, November 23, 1959; "Scientists Here Find Method to Take Strontium from Milk," *St. Louis Post-Dispatch*, June 24, 1960; Box 30, Scrapbooks, CEI Records, SL 69, WHMC, UMSL; "St. Louis Dairy Seeking a Way to Take Strontium Out of Milk," *New York Times*, July 3, 1960, p. 25.

48. Mrs. Phyllis (Kenneth L.) Smith to Orville Freeman, Secretary of Agriculture and Audrey (Mrs. William E.) Elder to Senator Estes Kefauver, February 5, 1962, box 3748, RG 16, SOA, GC, 1906–76, NARA; Edward Walsh to James J. Delaney, May 30, 1962, James J. Delaney Papers, Series 1, box 1, Special Archives and Collections, University at Albany, Albany, N.Y.

49. Stewart Huber, 1962 Annual Meeting Minutes, 43, Consolidated Badger Cooperative Records, box 3, Green Bay Mss 104, WHS, Madison, Wisc.; Ronald McCamus, Agricultural Extension Service, Kandiyohi County, Minn., to Orville Freeman, April 2, 1962, box 3748, RG 16, SOA, GC, 1906–76, NARA.

50. Tom Wicker, "President Toasts Milk with Milk," *New York Times*, January 24, 1962, p. 1; Tom Wicker, "President Hails American Cow: Again Pleads for Drinking Milk," *New York Times*, January 25, 1962, p. 11.

51. Robert Plumb, "Fallout Iodine Is Found in Milk," *New York Times*, June 21, 1959, p. 76; "Iodine-131 in Fallout: A Public Health Problem," *Consumer Reports* 27 (September 1962): 446–447.

52. Mrs. Reiser to Baby Tooth Survey committee, 1964, box 13, CEI Records, SL 69, WHMC, UMSL.

53. U.S. Department of Health, Education, and Welfare, Public Health Service and Food and Drug Administration, "Fallout Surveillance and Protection," October 26, 1961, in box 3748, RG 16, SOA, GC, 1906–76, NARA.

54. "Fallout Surveillance and Protection," 6.

55. Dr. Robert Barr, Minnesota Department of Health, to Dr. Donald Chadwick, Division of Radiological Health, August 27, 1962, box 3748, RG 16, SOA, CG, 1906–76, NARA.

56. Walter Mondale Brief, August 21, 1962; Willard Cochrane to Orville Freeman, August 16, 1962, box 3748, RG 16, SOA, GC, 1906–76, NARA; Russell Asleson, "Milk Producers Adopt Voluntary Anti-Fallout Plan," *Minneapolis Daily Tribune*, August 16, 1962, pp. 1, 11; Victor Cohn, "Why State Took Precautions to Curb Fallout Danger," *Minneapolis Sunday Tribune*, August 19, 1962, pp. 1, 5. Utah farmers took similar measures in the summer of 1962. See U.S. Department of Health, Education, and Welfare, August 17, 1962, box 3748, Folder Dairy Products, April 21–September 20 (1 of 2), RG 16, SOA, GC, 1906–76, NARA.

57. Quote from Henry Ackerman to Orville Freeman, September 1962. See also Theodore Maki, Hagen Metso, Carl Johnson, Harry Metso, Leonard Kauppila, and Raymond Rengo to Orville Freeman, August 22, 1962, box 3748, RG 16, SOA GC, 1906–76, NARA.

58. Asleson, "Milk Producers Adopt Voluntary Anti-Fallout Plan," 11.

59. Daily Fallout Report, September 12, 1962, box 3748, RG 16, SOA, GC, 1906–76, NARA.

60. Walter Mondale Brief, August 21, 1962, box 3748, RG 16, SOA, GC, 1906–76, NARA.

61. Carson to Bauer, November 12, 1963, box 30, Scrapbook 4, SL 69, WHMC, UMSL.

62. Investigational Program for Antibiotic and Pesticide Residues in Fluid Market Milk, August 1, 1958, box 2650, Chester Hubble to Mrs. Florence Latner, January 11, 1960, box 2820; T. W. Workman to M. R. Stephens, March 4, 1960, box 2821; all in FDA General Subject Files, RG 88, NARA.

63. Christopher Bosso, *Pesticides and Politics: The Life Cycle of a Public Issue* (Pittsburgh: University of Pittsburgh Press, 1987); Thomas Dunlap, *DDT: Scientists, Citizens, and Public Policy* (Princeton: Princeton University Press, 1981).

64. Rachel Carson's *Silent Spring,* while noting the hazards of tainted milk to consumers, did not address the economic impact on farmers. Carson, *Silent Spring* (Boston: Houghton-Mifflin, 1962), 159–160; Joshua Blue Buhs, "Dead Cows on a Georgia Field: Mapping the Cultural Landscape of the Post–World War II American Pesticide Controversies," *Environmental History* 7 (January 2002): 104, 108–109.

65. R. N. Annend, "DDT NO Panacea," *Hoard's Dairyman* 89, no. 19 (October 10, 1944): 541; similar findings reported in Valerie Guneter and Craig Harris, "Noisy Winter: The DDT Controversy in the Years before *Silent Spring*," *Rural Sociology* 63 (June 1998): 179–180.

66. Evert Wallenfeldt, "Producing Quality Milk on the Farm," *Hoard's Dairyman* 92, no. 8 (January 25, 1947): 53.

67. "Sweet Spring…and Flies Again!" *Hoard's Dairyman* 92, no. 8 (April 25, 1947): 376; "Experience at West Virginia University with DDT," *Hoard's Dairyman* 92, no. 3 (February 10, 1947): 110; Charles White, "Controlling Insects on the Farm," *Milk Dealer* 38, no. 3 (December 1948): 112–113; Edmund Russell, *War and Nature: Fighting Humans and Insects with Chemicals from World War I to Silent Spring* (New York: Cambridge University Press, 2001), 158–159.

68. "Flies and Fly Control," *Hoard's Dairyman* 77 (June 10, 1932): 289, 298; G. L. Nelson, "Milking Barns for the South and Southwest," *Hoard's Dairyman* 95 (January 25, 1950): 41, 50; Linda Nash, *Inescapable Ecologies: A History of Environment, Disease, and Knowledge* (Berkeley: University of California Press, 2006), 129.

69. Dunlap, *DDT,* 63–65.

70. "Fly Control for 1950," *Hoard's Dairyman* 95, no. 8 (April 25, 1950): 307, 314; DuPont de Nemours and Company, Grasselli Chemicals Department, "How to Use DuPont Dairy Insecticides," Pamphlet Collection, Hagley Museum and Library, Wilmington, Del.

71. Robert Muck to Congressman Melvin Laird, September 20, 1961, box 3610, folder Insecticides, RG 16, SOA, GC 1906–76, NARA.

72. Inspector Edward Duke to Director, Minneapolis District, June 21, 1961, FDA General Subject Files, box 3026, RG 88, NARA.

73. Inspector Edward Duke to Director, Minneapolis District, June 21, 1961, FDA General Subject Files, box 3026, RG 88, NARA; Robert Muck to Melvin Laird, September 20, 1961, Frank Welch, assistant secretary of agriculture to Congressman Melvin Laird, October 9, 1961, box 3610, RG 16, SOA, GC 1906–75, NARA.

74. F. C. Bishopp to Hon. Carl Hayden, September 4, 1952, box 5, Correspondence and Records, Records of Bureau of Dairy Industry, 1944–1953, RG 152, NARA.

75. The flip-side of this logic was that any cases of adulteration were the result of improper chemical use, not the chemicals themselves.

76. Hays, *Beauty, Health, and Permanence*, 175.

77. Report of Roy Sandberg and J. J. Hanagan, inspectors to Director of the Detroit District, September 22, 1960, box 2822; memo of phone conversation between Mr. Henry Hortman, director of Division of Milk and Dairy Products, State Board of Health, New Orleans, Louisana, and E. C. Boudreaux, New Orleans District FDA, May 19, 1961, box 3025; Inspector M. J. Matlock to Director, Buffalo District, October 26, 1961, box 3026; Inspector Gordon E. Hanson to Director, Minneapolis District, July 27, 1961, box 3023, RG 88, FDA General Subject Files, NARA; Emanuel Sidel to Mr. Orville Freeman, August 3, 1966, box 4543, RG 16, SOA, GC, 1906–75, NARA.

78. Pete Daniel, *Toxic Drift: Pesticides and Health in the Post–War II South* (Baton Rouge: Louisiana State University Press, 2005), 14–47. A 1955 pesticide survey indicated the highest percentages of contaminated milk in the Atlanta, Los Angeles, New Orleans, and San Francisco districts. Investigational Program on Antibiotic and Pesticide Residues, August 1, 1958, box 2650, FDA General Subject Files, RG 88, NARA.

79. W. T. Smith to James O. Eastland, August 1, 1960, box 3455; see also George Mehren to Senator Tower, October 30, 1967, box 4644, RG 16, SOA, GC, 1906–75, NARA. For orchard-growing conditions, see Otto Bohnert, President, Jackson County Extension Agricultural Council, Medford, Oregon, to Honorable Wayne Morse, May 4, 1960, box 3455, RG 16, SOA, GC, 1906–75, NARA.

80. Louis Burrus to Senator John Tower, September 12, 1967; "DDT Residues in Milk—El Paso and Pecos Texas Areas," report dated April 28, 1967, box 4644, RG 16, SOA, GC, 1906–75, NARA.

81. Joe Spink, Borden Foods, to Senator John Stennie, May 4, 1966, box 3839, FDA General Subject Files, NARA.

82. Bosso, *Pesticides and Politics*, 67–69.

83. R. G. Lytle, Manager, United Dairymen of Arizona, to Mr. Gordon Wood, FDA Director May 4, 1964, box 3684, FDA General Subject Files, RG 88, NARA; David Vail, "'Kill That Thistle': Rogue Sprayers, Bootlegged Chemicals, and the Kansas Chemical Laws, 1950–1980," paper presented at the 2008 Agricultural History Society Conference, University of Nevada-Reno, June 19, 2008.

84. J. G. Rosier, Report on DDT Residues in Milk, El Paso and Pecos Texas Areas, box 4644; John Schnittker to J. W. Trimble, January 25, 1966, box 4543; both in SOA, GC, 1906–75, NARA.

85. Antibiotic and Pesticide Residues in Fluid Market Milk, August 1, 1958, box 2650; Roy D. Sandberg and J. J. Hanagan to Director Detroit District, September 22, 1960, box 2822; both in RG 88, FDA General Subject Files, NARA.

86. R. G. Lytle, United Dairymen of Arizona, to Mr. Gordon Wood, May 4, 1964, box 3684, FDA General Subject Files, RG 88, NARA.

87. A. C. Pollard to Senator Engle, May 13, 1960, box 2821, FDA General Subject Files, RG 88, NARA.

88. Wesley Sawyer to Hon. Oren Harris, March 22, 1960, box 2821, FDA General Subject Files, RG 88, NARA.

89. Lawrence Cameron, Miller Chemical and Fertilizer Corporation, to Orville Freeman, July 20, 1964, box 4132; Resolutions Adopted by American Farm Bureau Federation, Decem-

ber 10, 1964, box 4360; both in SOA, GC, 1906–75, NARA; Interim Report: Investigations of Pesticide Residues in Milk and Milk Products, box 3025, FDA General Subject Files, RG 88, NARA.

90. Daniel, *Toxic Drift*, 132–133; Bosso, *Pesticides and Politics*, 124–125; "Comments on the Recommendations in Report on 'No Residue' and 'Zero Tolerance' issued in June 1965 by NAS-NRC Pesticide Residues Committee," June 1965, box 4359, SOA, GC, 1906–75, NARA.

91. John Harvey, Deputy Commissioner, FDA, to Secretary, April 15, 1960, box 2821; Frank Vorhes, "Pesticide Residues in Milk," to Insecticide Society of Washington, April 17, 1957, box 2307; Irvan Kerlan, Associate Medical Director, Bureau of Medicine, to Samuel Fomon, M.D., box 3026; all in FDA General Subject Files, RG 88, NARA.

92. Mrs. Ruth Desmond to Maryland and Virginia Milk Producers, May 8, 1964, Anna Naragon to Senator Warren Magnuson, May 22, 1964; both in box 4109, SOA, GC, 1906–75, NARA.

93. Diane Tempest, Geralyn Tatro, James Johannek, Candace Ryan, Carol Klein, Joan Fischback, Frederick Fruci, Jennifer Heimer, Thomas Adam, Patricia Olson, Jeanne Weber, Dianne Henschell, and Sheila Moharty to Secretary of Agriculture Orville Freeman, all dated February 23, 1966, box 4543; Doris Honig to Hon. Frank Thompson, November 13, 1965, box 4358, SOA, GC 1906–75, NARA.

94. Mary Douglas, *Purity and Danger: An Analysis of the Concepts of Pollution and Taboo* (London: Routledge, 1966), 2–4.

Epilogue

1. Dale Bauman, "Bovine Somatotropin: Review of an Emerging Animal Technology," *Journal of Dairy Science* 75 (1992): 3435; R. Collier, "Regulation of rBST in the United States," 3 *AgBioForum* (2000): 2–3; William Serrin, "Cornell Reports Big Milk Production Gains," *New York Times*, October 19, 1985; Keith Schneider, "Biotech's Cash Cow," *New York Times*, June 12, 1988; Keith Schneider, "Betting the Farm on Biotech," *New York Times*, June 10, 1990; Frederick Buttel, "Nature's Place in the Technological Transformation of Agriculture: Some Reflections on the Recombinant BST Controversy in the USA," *Environment and Planning A* (1998): 1156.

2. Bauman, "Bovine Somatotropin," 3432; U.S. Congress, Office of Technology Assessment, *U.S. Dairy Industry, Biotechnology, and Policy Choices—Special Report*, OTA-F-470 (Washington, D.C.: Government Printing Office, May 1991), 35–37, 43; Henry Miller, "Don't Cry over rBST Milk," *New York Times*, June 29, 2007.

3. Gina Kolata, "When Geneticists' Fingers Get in the Food," *New York Times*, February 20, 1994.

4. Bruce Ingersoll, "FDA, Prodded by Independent Surveys, Begins Testing Milk for Animal Drugs," *Wall Street Journal*, January 2, 1990; Bruce Ingersoll, "New York Milk Supply Highly Tainted, TV Station Says, Based on Own Survey," *Wall Street Journal*, February 8, 1990; Bruce Ingersoll, "FDA Plans a Nationwide Test of Milk for Antibiotics, Other Drug Residues," *Wall Street Journal*, December 28, 1990.

5. For mastitis studies, see T. C. White et al., "Clinical Mastitis in Cows Treated with Sometribove (Recombinant Bovine Somatotropin) and Its Relationship to Milk Yield," *Journal of Dairy Science* 77 (August 1994): 2249–2260; L. J. Judge, R. J. Erskine, and P. C. Bartlett, "Recombinant Bovine Somatotropin and Clinical Mastitis: Incidence, Discarded Milk Following Therapy, and Culling," *Journal of Dairy Science* 80 (December 1997): 3212–3218; Kolata, "When Geneticists' Fingers Get in the Food"; Keith Schneider, "Despite Critics, Dairy Farmers Increase Use of Growth Hormone," *New York Times*, October 30, 1994. For an insightful analysis of the shifting rhetoric of anti-rBGH activists, see Frederick Buttel,

"The Recombinant BGH Controversy in the United States: Toward a New Consumption Politics of Food?" *Agriculture and Human Values* 17 (2000): 16.

6. Ultimately, while nearly 10 percent of the nation's dairy farmers exited the industry through the program, milk production declined only a fraction of 1 percent, for those remaining in the industry continued to boost production. Elizabeth Way, "The Lowing Herds Wind Slowly Over the Hill," *New York Times*, March 8, 1986; Keith Schneider, "In Dairy State, A Way of Life Ends for Many: Slaughter of Cows Seen by Some as Solution," *New York Times*, April 2, 1986; "Agency Sending Doomed Cattle to the Needy," *New York Times*, December 26, 1986; "Milk Program Assessed," *New York Times* June 9, 1988; Frederick Buttel, "Agricultural Research and Farm Structural Change: Bovine Growth Hormone and Beyond," *Agriculture and Human Values* 3 (Fall 1986): 89–92.

7. Robert Kalter, "The New Biotech Agriculture: Unforseen Economic Consequences," *Issues in Science and Technology* 2 (Fall 1985): 128–131; Ward Sinclair, "Biotechnology and the Milk Glut: New Growth Hormone Raises Uproar in Dairy Industry," *Washington Post*, June 23, 1986; Buttel, "The Recombinant BGH Controversy in the United States," 5.

8. David Berry, "Effects of Urbanization on Agricultural Activities," *Growth and Change* (July 1978): 5; Howard Conklin and Richard Dymsza, "Maintaining Viable Agriculture in Areas of Urban Expansion" (Albany: New York State Office of Planning Services, 1972), 35, 44.

9. Tim Lehman, *Public Values, Private Lands: Farmland Preservation Policy, 1933–1985* (Chapel Hill: University of North Carolina Press, 1995), 107; Christopher McGrory Klyza and Stephen Trombulak, *The Story of Vermont: A Natural and Cultural History* (Hanover, N.H.: University of New England Press, 1999), 115–129.

10. Ralph Heimlich and Douglas Brooks, "Metropolitan Growth and Agriculture: Farming in the City's Shadow," Agricultural Economic Report 619 (Washington, D.C.: Economic Research Service, USDA, 1989), 15.

11. Adam Rome, *Bulldozer in the Countryside: Suburban Sprawl and the Rise of American Environmentalism* (Cambridge: Cambridge University Press, 2001), 138–152.

12. Lehman, *Public Values, Private Lands*, 100–102, 123–132; Klyza and Trombulak, *The Story of Vermont*, 122–123; Rome, *Bulldozer in the Countryside*, 240–253, 266.

13. Schneider, "Vermont Resists Some Progress in Dairying," *New York Times*, August 27, 1989; Susan Bentley and William Saupe, "Exits from Farming in Southwest Wisconsin, 1982–1986," Agricultural Economic Report 631 (Washington, D.C.: Economic Research Service, USDA, 1990), 3; William Browne and Larry Hamm, "Political Choice and Social Values: The Case of bGH," *Policy Studies Journal* 17 (Fall 1988): 183–184; Keith Schneider, "A Drug for Cows Churns up Opposition," *New York Times*, May 20, 1990.

14. Kevin Kelley, "Condos Where Cows Once Grazed," *Progressive* 53 (November 1989): 28; Robert Kalter, "The New Biotech Agriculture: Unforeseen Economic Consequences," *Issues in Science and Technology* 2 (Fall 1985): 128–131; Ward Sinclair, "Biotechnology and the Milk Glut: New Growth Hormone Raises Uproar in Dairy Industry," *Washington Post*, June 23, 1986.

15. Donna Walters, "Two States OK Milk Hormone Moratoriums," *Los Angeles Times*, April 28, 1990.

16. Patricia Strach, *All in the Family: The Private Roots of American Public Policy* (Stanford: Stanford University Press, 2007), 142–144.

17. Warren J. Belasco, *Appetite for Change: How the Counterculture Took on the Food Industry*, 2nd ed. (Ithaca: Cornell University Press, 2003), 142–144; Alissa Hamilton, *Squeezed: What You Don't Know about Orange Juice* (New Haven: Yale University Press, 2009), 31–103.

18. Keith Schneider, "Lines Drawn in a Battle over a Milk Hormone," *New York Times*, March 9, 1994; Karen Cohen, "BGH Testimony: FDA Hears Milk Labeling Arguments," *Wisconsin State Journal*, May 7, 1993; Robert Steyer, "Label Rule on BST Unpopular: FDA Makes It Hard to Prove a Negative," *St. Louis Post Dispatch*, February 13, 1994; Diane

Gershon, "FDA Panel Sees Problems with Labeling of Milk Hormone," *Nature* 363 (May 13, 1993): 107.

19. Lawrence Glickman, *Buying Power: A History of Consumer Activism in America* (Chicago: University of Chicago Press, 2009), 61–89, 175–188.

20. Marian Burros, "More Milk, More Confusion: What Should the Label Say?" *New York Times*, May 18, 1994.

21. Schneider, "Lines Drawn in a Battle over a Milk Hormone"; David Barboza, "Monsanto Sues Dairy in Maine over Label's Remarks on Hormones," *New York Times*, July 12, 2003; Allyce Bess, "Got Posilac?" *St. Louis Post Dispatch*, August 10, 2003.

22. Buttel, "The Recombinant BGH Controversy in the United States," 16–17, 19.

23. Andrew Martin, "Monsanto to Sell off Hormone Business," *New York Times*, August 7, 2008; Andrew Martin, "Fighting on a Battlefield the Size of a Milk Label," *New York Times*, March 9, 2008.

24. Buttel, "The Recombinant BGH Controversy in the United States," 16–17; Andrew Pollack, "Which Cows Do You Trust?" *New York Times*, October 7, 2006; Martin, "Fighting on a Battlefield the Size of a Milk Label."

25. Martin, "Monsanto to Sell off Hormone Business"; Martin, "Fighting on a Battlefield the Size of a Milk Label."

26. Belasco, *Appetite for Change*, 28, 68–76; Bruce Schulman, *The Seventies: The Great Shift in American Culture, Society, and Politics* (Cambridge: De Capo, 2001), 78–80.

27. Kate Murphy, "More Buyers Asking: Got Milk Without Chemicals?" *New York Times*, August 1, 1999; Carolyn Dimitri and Kathryn M. Venezia, "Retail and Consumer Aspects of the Organic Milk Market" (Washington, D.C.: Economic Research Service, U.S. Department of Agriculture, May 2007), 2.

28. Matt Mariola, Michelle Miller, and John Hendrickson, *Organic Agriculture in Wisconsin: 2003 Status Report* (Madison: Center for Integrated Agricultural Systems, University of Wisconsin-Madison, 2003), 1.

29. Catherine Greene, et. al. *Emerging Issues in the U.S. Organic Industry* (Washington, DC: Economic Research Service, U.S. Department of Agriculture, 2009), 10–12. Georgina Gustin, "Market Goes Sour for Organic Milk," *St. Louis Post Dispatch*, July 3, 2009; Edward Janus, *Creating Dairyland* (Madison: University of Wisconsin Press, 2011), 132–133.

30. Belasco, *Appetite for Change*, 93; Dimitri and Venezia, "Retail and Consumer Aspects of the Organic Milk Market," 6–7.

31. Melanie Warner, "Wal-Mart Eyes Organic Food, and Brand Names Get in Line," *New York Times*, May 12, 2006.

32. Melanie Warner, "A Milk War Over More Than Price," *New York Times*, September 16, 2006; Michael Pollan, "The Organic-Industrial Complex," *New York Times Magazine* (May 13, 2001): 30–37, 57–58, 63, 65; Andrew Martin, "Critics Say Dairy Tests the Boundaries and Spirit of What 'Organic' Means," *Chicago Tribune*, August 20, 2006.

33. Pollan, "The Organic-Industrial Complex," 30–37, 57–58, 63, 65; Melanie Warner, "What Is Organic? Powerful Players Want a Say" *New York Times*, November 1, 2005; Warner, "A Milk War Over More Than Price."

34. U.S. Department of Agriculture, National Organic Program Online Training Slides, "National Organic Program—Access to Pasture," March 26, 2010, accessed March 10, 2012, at http://www.ams.usda.gov/AMSv1.0/getfile?dDocName=STELPRDC5083618.

35. Kelliann Blazek et al., *Organic Agriculture in Wisconsin: 2009 Status Report* (Center for Integrated Agricultural Systems, University of Wisconsin-Madison, 2010), 11, 13–15.

36. Marcia Headrick et al., "The Epidemiology of Raw Milk–Associated Foodborne Disease Outbreaks Reported in the United States, 1973 through 1992," *American Journal of Public Health* 88 (August 1998): 1220.

37. William Greer, "Raw Milk Is Curbed," *New York Times*, January 7, 1987.

38. Nina Planck, *Real Food: What to Eat and Why* (New York: Bloomsbury, 2006), 67–68; David Gumpert, *The Raw Milk Revolution: Behind America's Emerging Battle over Food Rights* (White River Junction, Vt.: Chelsea Green, 2009), 87; Sally Fallon Morell, "Got Milk? Raw Milk May Be Illegal Locally, but Fans Know Which Barn Doors to Knock on to Find It," *Washington Post*, June 5, 2002.

39. Belasco, *Appetite for Change*, 37–42.

40. Planck, *Real Food*, 67–68.

41. "How We Use Our Pastures," *Hoard's Dairyman* 113 (November 10, 1968): 1248–1249, 1271, 1273.

42. Sam Hodges, "Got Raw Milk?" *Dallas Morning News*, February 21, 2010.

43. Alden Manchester and Don Blayney, *Milk Pricing in the United States*, Bulletin 761 (Washington, D.C.: U.S. Department of Agriculture, 2001), 13–14; Anthony DePalma, "Dairies Fewer but Stronger," *New York Times*, August 16, 1986.

44. Matthew DeFour, "Raw All Right?" *Wisconsin State Journal*, December 16, 2009; see other stories of economic motivations in Gumpert, *Raw Milk Revolution*, 156–157.

45. Oregon Department of Agriculture Food Safety Division, "Dairy Program," accessed October 15, 2010, at http://www.oregon.gov/ODA/FSD/program_dairy.shtml.

46. Gumpert, *Raw Milk Revolution*, xxv–xxix.

47. Salatin, foreword to Gumpert, *Raw Milk Revolution*, xi.

48. Glickman, *Buying Power*, 276–279, 290–294; Kirk, *Counterculture Green*, 183–184; Steinberg, "Can Capitalism Save the Planet?" 7–11.

49. J. M. Ketler, "Campylobacter and Helicobacter," in *Medical Microbiology: A Guide to Microbial Infections: Pathogenesis, Immunity, Laboratory Diagnosis, and Control*, ed. David Greenwood et al., 300–304 (Edinburgh: Churchill Livingstone, 2007); S. Benson Werner et al., "Association between Raw Milk and Human Salmonella Dublin Infection," *British Medical Journal* 2 (July 1979): 238–241; Joshua Fierer, "Invasive *Salmonella Dublin* Infections Associated with Drinking Raw Milk," *Western Journal of Medicine* 38 (May 1983): 665–669; H. Anderson et al., "Salmonella Dublin Associated with Raw Milk—Washington State," *MMWR* 30 (August 7, 1981): 373–374; Morris E. Potter, "Unpasteurized Milk: The Hazards of a Health Fetish," *Journal of the American Medical Association* 252 (October 1984): 2050; Patrick McDonough et al., "*Salmonella Enterica* Serotype Dublin Infection: An Emerging Infectious Disease for the Northeastern United States," *Journal of Clinical Microbiology* 37 (August 1999): 2418.

50. Julie Deardorff and Tim Jones, "Milk in the Raw: 'Rights' and Risk," *Los Angeles Times*, May 17, 2010; P. J. Huffstutter, "In Fight over Raw Foods, Choice and Safety Clash," *Los Angeles Times*, July 25, 2010; Planck, *Real Food*, 76–82; Gumpert, *Raw Milk Revolution*, 52, 81–98; Michael Bell and Ozlem Altiok, "Perceptions of Raw Milk's Risks and Benefits," Research Brief No. 83 (Madison, Wisc.: Center for Integrated Agricultural Systems, June 2010): 1.

51. DeFour, "Raw All Right?".

52. California Department of Food and Agriculture, "Bovine TB: California Update," October 2009, available at http://www.cdfa.ca.gov/ahfss/Animal_Health/pdfs/TB/Bovine_TB_Update_100109.pdf.

53. Daniel O'Brien et al., "Managing the Wildlife Reservoir of *Mycobacterium bovis*: The Michigan, USA, Experience," *Veterinary Microbiology* 112 (2006): 313–323; Eric Sharp, "Michigan's Struggle with Bovine TB," *Outdoor Life* (September 2001): 59.

SELECTED BIBLIOGRAPHY

Primary Sources

MANUSCRIPT COLLECTIONS

Archives of the National Academy, Washington, D.C.
　Papers of the Committee on Food Habits, Division of Anthropology and Psychology, National
　　Research Council
　Records of the National Research Council, Biology and Agriculture Division, Committee on
　　Food and Nutrition
Archives Center, National Museum of American History, Smithsonian Institution, Washington, D.C.
　Carvel Ice Cream Records, 1934–1989
　Eskimo Pie Corporation Records, 1921–1996
　Frances S. Baker Product Cookbooks, 1900–1990
　Good Humor Ice Cream Collection
　Louisan E. Mamer Rural Electrification Administration Papers, 1927–1994
　N. W. Ayer Advertising Agency Records
　Sam de Vincent Collection of Illustrated Sheet Music, 1790–1987
　Southern Agriculture Oral History Project Records
　Warshaw Collection of Business Americana, ca. 1790–1945
California State Parks Archive, Sacramento, California
　California Dairy Industry History Collection, 1856–1986
Library of Congress, Washington, D.C.
　America Eats Papers, Notes, Reports, and Essays, Records of the U.S. Work Projects
　　Administration
Minnesota Historical Society, St. Paul, Minnesota
　A. J. and Marie McGuire Papers
National Archives and Records Administration, College Park, Maryland
　Records of the Bureau of Dairy Industry, General Correspondence, 1944–1953. Record
　　Group 152
　Records of the Food and Drug Administration, General Subject Files, 1938–1974. Record
　　Group 88
　Records of the Office of the Secretary of Agriculture, General Correspondence, 1906–1970.
　　Record Group 16
　Records of the Office of the Secretary of Agriculture, Historical File, 1900–1959. Record
　　Group 16-G
Manuscripts and Special Collections, New York State Library, Albany, New York
　Boyd Sherwood Diaries, 1927–1955

Elihu Gifford Papers, 1888–1931
Shields Library, University of California-Davis, Department of Special Collections, Davis, California
Sunshine Farm Dairy Papers, 1945–1980
University of Wisconsin Archives, Steenbock Library, Madison, Wisconsin
Office of the Dean and Director; General Files, University of Wisconsin College of Agriculture.
H. L. Russell
Vermont Historical Society, Barre, Vermont
Mallary Farm Records, 1935–1988
Virginia Historical Society, Richmond, Virginia
Franklin Pope Wilson Papers, 1907–1928
Hollerith Family Papers, 1904–1985
Richmond Ice and Milk Mission, 1911–1928
Western Historical Manuscripts Collection, St. Louis, University of Missouri–St. Louis
Committee for Environmental Information Records
St. Louis Consumer Federation Papers, 1935–1977
Wisconsin Historical Society, Madison, Wisconsin
A-G Cooperative Creamery Papers, 1923–1974
Consolidated Badger Cooperative Records, 1931–1981. Green Bay Area Research Center
George Nelson Papers, Correspondence, 1908–1920
Otto Hunziker Papers, 1855–1950
Wisconsin Buttermakers' Association, Records, 1902–1950
Wisconsin Dairies Cooperative, Records, 1903–1965
Wisconsin Dairy and Food Division, Milk Adulteration Prosecution Reports, 1909–1950
Collection in possession of the author
L. Dale and Veleanor R. Beall Farm records

JOURNALS

Creamery and Milk Plant Monthly, alternate years, 1912–1930
Glass Lining: An Equipment and Service Magazine for the Dairy and Food Industries, 1928–1944
Hoard's Dairyman, alternate years, 1900–1970
Housewives' League Magazine
Ice Cream Field, 1945–1960
Ice Cream Trade Journal, 1945–1960
Modern Plastics, 1929–1934
News Bulletin, National Farm Chemurgic Council, 1940–1941
Pediatrics, 1900–1915
Plastics: Periodical Devoted to the Manufacture and Use of Composition Products, 1929–1930
Proceedings of the American Dry Milk Institute, 1931–1947

ORAL HISTORY INTERVIEWS

Blochowiak, B. L., interview by Dale E. Treleven, June 10, 1977. Wisconsin Agriculturalists' Oral History Project. Wisconsin Historical Society.
Eyrman, Elsie Brawner, interview by Laurie Mercier, September 28, 1982. Montana Historical Society, Helena, Montana.
Ihde, Aaron, interview by William K. Laderfer, May 15, 1963. University of Wisconsin-Madison Archives, Madison, Wisconsin.
Lucia, Floyd, interview by Dale Treleven, January 16, 1976. Wisconsin Agriculturalists' Oral History Project, Interview M75–34. Wisconsin Historical Society.
Myers, D. Wayne, interview by Rex C. Myers, September 14, 1982. Montanans at Work Oral History Series, Montana Historical Society.
Peltier, Victor, and Marie Peltier, interview by Laurie Mercier, September 22, 1983. Montana Historical Society.
Schurch, Walter E., interview by Lois McDonald, August 23, 1983. Montana Historical Society.

Segerstrom, Margaret, and Rangar Segerstrom, interview by Dale E. Treleven, September 29, 1976. Wisconsin Agriculturalists' Oral History Project, Wisconsin Historical Society.

Sullivan, Elizabeth, interview by Dale E. Treleven, May 11, 1977. Wisconsin Historical Society.

BOOKS AND ARTICLES

Abbott, J. S. "The Food and Economics of Skim Milk." *American Journal of Public Health* 30 (March 1940): 237–239.

Adams, Walter. "The 100% American Way to Die." *Better Homes and Gardens* 26 (November 1947): 32, 224–230.

Albrecht, Arthur E. "Food: Its Journey to Your Table." *Good Housekeeping* 80 (February 1920): 73, 136, 139.

Albright, J. L., S. L. Tuckey, and G. T. Woods. "Antibiotics in Milk: A Review." *Journal of Dairy Science* 44, no. 5 (May 1961): 779–807.

Alsberg, C. L. "Epidemics Attributable to Pasteurized Milk." *American Journal of Public Health* 13 (March 1923): 246.

Alvord, Henry. "Modern Dairy." *Current Literature* 29 (November 1900): 598.

———. "Utilization of By-Products of the Dairy." In United States Department of Agriculture, *Yearbook of Agriculture.* Washington, D.C.: Government Printing Office, 1898.

American Dried Milk Institute. *1953 Census of Dry Milk Distribution and Production Trends.* Chicago: American Dried Milk Institute, 1954.

Baker, S. Josephine. *Fighting for Life.* New York: Macmillan, 1939.

Barre, H. J., and L. L. Sammet. *Farm Structures.* New York: John Wiley, 1950.

Barrett, George Robert. "Time of Insemination and Conception Rate in Dairy Cattle." Ph.D. diss., University of Wisconsin-Madison, 1948.

Bartlett, C. J. "Does Commercial Pasteurization Destroy Tubercle Bacilli in Milk?" *American Journal of Public Health* 13 (October 1923): 807–809.

Bass, Stephen. *Plastics and You.* N.p.: Eastwood-Steli Company, 1947.

Benkendorf, G. H., and K. L. Hatch. *Profitable Dairying: A Manual for Farmers and Dairymen.* Privately printed: K. L. Hatch, 1908.

Carson, Rachel. *Silent Spring.* Boston: Houghton Mifflin, 1962.

Cavanaugh, J. F., Secretary. Purebred Dairy Cattle Association "Remarks." *Proceedings of Fifth Annual Convention of the National Association of Artificial Breeders.* Springfield, Mo.: National Association of Artificial Breeders, 1952.

Cherry-Bassett Company. *Complete Catalog of Equipment for Milk Plants, Dairies, and Creameries.* Baltimore: Cherry-Bassett Company, 1921.

———. *Condensed Catalog and Price Lists of Dairymens Supplies and Equipment.* Chicago: Cherry-Burrell Corporation, 1939.

Clark, Newton. "Butter Prices, From Farm to Consumer." *Bulletin of the U.S. Bureau of Labor Statistics.* Washington, D.C.: Government Printing Office, 1915.

Claybaugh, G. A., and F. E. Nelson. "The Effect of Antibiotics in Milk: A Review." *Journal of Milk and Food Technology* 14 (September–October 1951): 155–164.

Collins, J. H. "The Problem of Drugs for Food-Producing Animals and Poultry." *Food, Drug, Cosmetic Law Journal* 6, no. 11 (November 1951): 876–877.

Commoner, Barry. "The Fallout Problem." *Science* 127 (May 2, 1958): 1023–1026.

Cook, Hugh, and George Day. *The Dry Milk Industry: An Aid in the Utilization of the Food Constituents of Milk.* Chicago: American Dry Milk Institute, 1947.

Coolidge, Emelyn L. "Milk and Its Care in Warm Weather." *Ladies Home Journal* 34 (June 1917): 75.

Courter, Robert, and Mildred Galton. "Animal Staphylococcal Infections and Their Public Health Significance." *American Journal of Public Health* 52, no. 11 (November 1962): 1818–1827.

Cowden, Joseph. *Comparing Bulk and Can Milk Hauling Costs.* Farm Cooperative Service Circular 14. Washington, D.C.: United States Department of Agriculture, 1956.

Creamery Package Manufacturing Company. *Equipment and Supplies for Creameries, Milk Plants, Ice Cream Plants, Cheese Factories, Condensed and Evaporated Milk Plants, and Refrigerating Machinery.* Chicago: Creamery Package Manufacturing Company, 1923.

———. *The Handy Book of Dairy Supplies, Catalog No. 42.* Chicago: Creamery Package Manufacturing Company, 1942. Hagley Museum and Library, Trade Catalog Collection.

Crumbine, Samuel, and James Tobey. *The Most Nearly Perfect Food: The Story of Milk.* Baltimore: Williams and Wilkins, 1929.

Dahlberg, A. C., H. S. Adams, and M. E. Held. *Sanitary Milk Control and Its Relation to the Sanitary, Nutritive, and Other Qualities of Milk.* Publication 250. Washington, D.C.: National Academy of Sciences—National Research Council, 1953.

Educational Advertising Institute. *Chicago: A Century of Progress.* Chicago: Educational Advertising Institute, 1933.

E. I. DuPont de Nemours and Company. *Design for Selling: A Study of Impulse Buying.* Wilmington, Del.: E. I. DuPont de Nemours, 1946. Hagley Museum and Library, Pamphlet Collection.

Emerson, Haven. "Per Capita Milk Consumption from the Point of View of the Public Health Officer." *American Journal of Public Health* 14, no. 4 (April 1924): 291–292.

"Fallout in Our Milk, A Follow-Up Report." *Consumer Reports* 25 (February 1960): 64–70.

Farmer, Fannie Merritt. *Food and Cookery for the Sick and Convalescent.* Boston: Little, Brown, 1904.

"Fats in Diet: The Real Key to Heart Disease?" *Good Housekeeping* 155 (October 1962): 155–157.

Fistere, Charles. "Pesticide Residues: Legal Aspects." *Food, Drug, and Cosmetic Law Journal* 20 (December 1965): 684–694.

Flannery, James Joseph. "Water Pollution Control: Development of State and National Policy." Ph.D. diss., University of Wisconsin, 1956.

Foster, Wade. "Chemical Treatment of Dairy Wastes at Bad Axe, Mich.," *American City* 39 (September 1928): 122–123.

Fowler, John, ed. *Fallout: A Study of Superbombs, Strontium-90, and Survival.* New York: Basic, 1960.

Frandsen, Jules, and D. Horace Nelson. *Ice Creams and Other Frozen Desserts.* Amherst, Mass.: J. H. Frandsen, 1950.

Frank, Leslie. "The Unification of Milk Control in the United States." *American Journal of Public Health* 15 (November 1925): 977–980.

"Getting the Strontium-90 Out of the Milk." *Consumer Reports* 27 (February 1962): 63.

Guthrie, Edward Sewall. *The Book of Butter: A Text on the Nature, Manufacture, and Marketing of the Product.* New York: Macmillan, 1918.

Hale, William. *The Farm Chemurgic: Farmward the Star of Destiny Lights Our Way.* Boston: Stratford Company, 1934.

Hambridge, Gove. "The Grocery Revolution." *Ladies Home Journal* 45 (November 1928): 99, 119, 121.

Hansen, Chr. *Commercial Buttermilk or Cultured Milk: Principles and Practice of Preparation.* Little Falls, N.Y.: Chr. Hansen's Laboratory, 1915.

Harvey, John. "What Can Industry Do to Help Solve Problems of Antibiotics and Insecticides?" *Food, Drug, and Cosmetic Law Journal* 15 (April 1960): 256–262.

Hawk, Philip E. "What We Eat: And What Happens to It: The Truth about the Milk We Drink!" *Ladies Home Journal* 4 (February 1917): 29.

Hayward, Walter, and Percival White. *Chain Stores: Their Management and Operation.* 2nd ed. New York: McGraw-Hill, 1925.

Heineman, H. E., and C. B. Miller. "Pesticide Residues and the Dairy Industry." *Journal of Dairy Science* 44 (September 1961): 1775–1781.

Henry, W. A. "Comparative Value of Feeds, with Tables Giving Their Percentage of Digestible Nutrients." In *Creamery Patron's Handbook*, 31–43. Chicago: National Dairy Union, 1902.

Herman, H. A., and F. W. Madden. *The Artificial Insemination of Dairy Cattle: A Handbook and Laboratory Manual.* Columbia, Mo.: Lucas Brothers, 1947.

212 Selected Bibliography

Howard, Martha Crampton. "The Margarine Industry in the United States: Its Development under Legislative Control." Ph.D. diss., University of Wisconsin-Madison, 1951.

Hunziker, Otto. *The Butter Industry*. 3rd ed. LaGrange, Ill.: Hunziker, 1940.

Hussmann Refrigeration, Inc. *Self Service: The Sure Road to Greater Profits in Refrigerated Meats, Produce, Dairy Products, Frozen Foods*. St. Louis: Hussmann Refrigeration, 1945.

"Iodine-131 in Fallout: A Public Health Problem." *Consumer Reports* 27 (September 1962): 446–447.

Isaksson, Nils-Ivar. "Economies of Scale in Milk Production and the Competitive Position of the Family Farm." Master's thesis, University of Wisconsin, 1961.

Jester, William, William Wright, and Henry Welch. "Antibiotics in Fluid Milk: Fourth Nationwide Survey." *Antibiotics and Chemotherapy* 9 (July 1959): 393–397.

Joliffe, Norman. "Fats, Cholesterol, and Coronary Heart Disease." *Circulation* 20 (July 1959): 109–127.

Katznelson, H., and E. G. Hood. "Influence of Penicillin and Other Antibiotics on Lactic Streptococci in Starter Cultures Used in Cheddar Cheesemaking." *Journal of Dairy Science* 32, no. 11 (November 1949): 961–968.

Keys, Ancel. "Human Atherosclerosis and the Diet." *Circulation* 5 (1952): 115–118.

Kosikowski, F. V. "An Inquiring Look at Antibiotic Residue Control." *Journal of Dairy Science* 44 (August 1961): 1554–1557.

Larson, B. L., and K. E. Ebner. "Significance of Strontium-90 in Milk: A Review." *Journal of Dairy Science* 41 (December 1959): 1651–1656.

Levin, William. "The Relation of the State to a Sanitary Milk Supply." *American Journal of Public Health* 13 (July 1923): 557–561.

Libby, W. F. "Radioactive Fallout." *Proceedings of the National Academy of Sciences of the United States of America* 44 (August 1958): 800–820.

———. "Radioactive Fallout and Radioactive Strontium." *Science* 123 (April 20, 1956): 657–660.

Louden Machinery Company. *Some Interesting Facts on a Homely Subject*. Fairfield, Iowa: Louden Machinery Company, 1907.

Lough, S. Allan. "Fallout and Our Food Supply." *Food, Drug, and Cosmetic Law Journal* 117 (February 1962): 117.

Macklin, Theodore. "A History of the Organization of Creameries and Cheese Factories in the United States." Ph.D. diss., University of Wisconsin-Madison, 1917.

MacNutt, Joseph Scott. *The Modern Milk Problem in Sanitation, Economics and Agriculture*. New York: Macmillan, 1917.

McCollum, E. V., and Nina Simmonds. *The Newer Knowledge of Nutrition: The Use of Foods for the Preservation of Vitality and Health*. New York: Macmillan, 1925.

McKay, G. L. *Facts about Butter: Suggestions for a Standard*. Bulletin 12. Chicago: American Creamery Butter Manufacturers, June 1918.

McKay, G. L., and C. Larsen. *Principles and Practice of Butter-making*. 2nd ed. New York: Wiley, 1908.

Merchant, Ival, and R. Allen Packer. *Handbook for the Etiology, Diagnosis, and Control of Infectious Bovine Mastitis*. 2nd ed. Minneapolis: Burgess, 1952.

Mercier, Ernest, and G. W. Salisbury. "Fertility Level in Artificial Breeding Associated with Season, Hours of Daylight, and the Age of Cattle." *Journal of Dairy Science* 30 (November 1947): 817–826.

Merrell-Soule Company. *Merrell-Soule Products: Powdered Milk and None Such Mince Meat*. Syracuse, N.Y.: Merrell-Soule Company, 1919.

"The Milk All of Us Drink—and Fallout." *Consumer Reports* 24 (March 1959): 102–103.

Moore, Marian, and Julia Pond. "Dessert Choices of Southern Urban Families." *Journal of Home Economics* 58 (October 1966): 659–661.

National Dairy Products Corporation. *New Ways with Ice Cream*. New York: Sealtest, 1949.

Neubauer, Loren, and Harry Walker. *Farm Building Design*. Englewood Cliffs, N.J.: Prentice Hall, 1961.

Noorlander, Daniel. *Milking Machines and Mastitis*. Compton, Calif.: Compton, 1960.

North, Charles Edward. *Farmers' Clean Milk Book*. New York: John Wiley, 1918.

Parker, Horatio Newman. *City Milk Supply*. New York: McGraw-Hill, 1917.

Paterson Parchment Paper Company. *Better Butter*. Passiac, N.J.: Paterson Parchment Paper Company, 1915. Hagley Museum and Library, Trade Catalog Collection, Wilmington, Del.

Perry, Enos J. *The Artificial Insemination of Farm Animals*. New Brunswick, N.J.: Rutgers University Press, 1945.

Peters, Lulu Hunt. *Diet for Children (and Adults) and the Kalorie Kids*. New York: Dodd, Mead, 1924.

Pou, John William. "A Study of Wisconsin Cooperative Artificial Insemination Associations." Master's thesis, University of Wisconsin-Madison, 1947.

Ravenel, Mazyck P. "Relations of Human and Bovine Tuberculosis." In *Sixth International Congress on Tuberculosis, Transactions. Volume 4, Part 2, Section 7: Tuberculosis in Animals and Its Relation to Man*, 683–687. Philadelphia: Fell Company, 1908.

Rockefeller, Mrs. John D. "Small Wayside Refreshment-Stand Competition." *Ladies Home Journal* 44 (November 1927): 30.

Rosenau, Milton J. *The Milk Question*. Boston: Houghton Mifflin, 1912.

Schaars, Marvin. "The Butter Industry of Wisconsin." Ph.D. diss., University of Wisconsin-Madison, 1932.

Shrader, J. H. "The Significance of the Tuberculin Test in a Pasteurized Milk Supply." *American Journal of Public Health* 15 (September 1925): 767–770.

Shumway, Henry. *A Hand-book on Tuberculosis among Cattle with Considerations of the Relation of the Disease to the Health and Life of the Human Family, and of the Facts Concerning the Use of the Tuberculin as a Diagnostic Test*. London: Sampson Low, Marston, 1895.

Spargo, John. *The Common Sense of the Milk Question*. New York: Macmillan, 1910.

"Special Report: Heart Attacks: Can Diet Prevent Them?" *Good Housekeeping* 151 (May 1961): 137–139.

Spencer, Steven. "Fallout: The Silent Killer, How Soon Is Too Late?" *Saturday Evening Post* 232 (September 29, 1959): 84.

Stanley, Louise, and Jessie Alice Cline. *Foods: Their Selection and Preparation*. Boston: Ginn and Company, 1935.

Stewart, Ethelbert. "Economics of Creamery Butter Consumption." *Monthly Labor Review* 21 (1925): 1–2.

Sutermeister, Edwin. *Casein and Its Industrial Applications*. New York: Chemical Catalog Company, 1927.

Sutermeister, Edwin, and Frederick Browne. *Casein and Its Industrial Applications*. New York: Reinhold Publishing, 1939.

Till, Irene. *Milk: The Politics of an Industry*. New York: McGraw-Hill, 1938.

Tobey, James, ed. *Public Health Law: A Manual of Law for Sanitarians*. Baltimore: Williams and Wilkins, 1926.

Todd, Frank. "Controlling Radioactive Fallout Contamination." *Journal of Dairy Science* 45 (December 1962): 1543–1557.

United Dairy Equipment Company. *Sales Manual and Instructions for Installation and Operation of the World-Famous Mechanical Cow*. West Chester, Penn.: United Dairy Equipment Company, 1943. Hagley Museum and Library, Trade Catalog Collection.

United States Steel Corporation. *Bulk Handling of Milk with Stainless Steel*. Pittsburgh: United States Steel Corporation, 1955. Hagley Museum and Library, Trade Catalog Collection.

Van Pelt, Hugh. *How to Feed the Dairy Cow: Breeding and Feeding Dairy Cattle*. Waterloo, Iowa: Fred Kimball Company, 1919.

Vermont Farm Machine Company. *United States Cream Separators*. Bellows Falls, Vt.: The Vermont Machine Company, 1910.

Vickers, H. R., L. Bagratuni, and Suzanne Alexander. "Dermatitis Caused by Penicillin in Milk." *Lancet* 159 (February 15, 1958): 351–352.

Ward, Archibald Robinson. *Pure Milk and the Public Health: A Manual of Milk and Dairy Inspection.* Ithaca, N.Y.: Taylor and Carpenter, 1909.

Weist, Edward. *The Butter Industry in the United States: An Economic Study of Butter and Oleomargarine.* New York: Columbia University Press, 1916.

Welch, Henry. "Control of Antibiotics in Food." *Food, Drug, and Cosmetic Law Journal* 12 (1957): 462–465.

Welch, H., William Jester, and J. M. Burton. "Antibiotics in Fluid Market Milk: Third Nationwide Study." *Antibiotics and Chemotherapy* 6 (May 1956): 369–374.

———. "Antibiotics in Fluid Milk." *Antibiotics and Chemotherapy* 5 (October 1955): 571–573.

Wesen, Donald Philip. "Studies on the Prevention and Treatment of Mastitis in Dairy Cows." Ph.D. diss., University of Wisconsin-Madison, 1969.

White, Paul Dudley and Curtis Mitchell, eds. *Fitness for the Whole Family.* Garden City, N.Y.: Doubleday, 1964.

"The Whole Milk Creamery Movement: What Is It? What Does It Mean? Where Will It Lead?" *National Butter and Cheese Journal* 28 (January 12, 1937): 10, 12, 14–15.

Wiley, Harvey W. "Milk." *Good Housekeeping* 35 (February 1918): 49–50, 137–138.

Wisconsin Cheese Makers' Association. *Proceedings of the Wisconsin Cheese Makers' Association Fifty-Sixth Annual Meeting, November 3–4, 1947.* Auditorium and Schroeder Hotel, Milwaukee, Wisconsin. Madison, Wisc.: Cantwell Press, 1913/1915–1954.

———. *Proceedings of Wisconsin Cheesemakers' Association Fifty-Seventh Annual Meeting and Centennial Convention, October 19–20, 1948.* Retlaw Hotel, Fond du Lac, Wisconsin. Madison, Wisc.: Cantwell Press, 1913/1915–1954.

———. *Proceedings of the Wisconsin Cheese Maker's Association Sixtieth Annual Meeting, October 24–25, 1951.* Retlaw Hotel and County Building, Fond du Lac, Wisconsin. Madison, Wisc.: Cantwell Press, 1913/1915–1954.

Wisconsin Dairymen's Association. *Thirty-Sixth Annual Report of the Wisconsin Dairymen's Association.* Madison, Wisc.: Democrat Printing Company, 1910.

Woll, F. W. *A Book on Silage.* Chicago: Rand, McNally, 1895.

GOVERNMENT DOCUMENTS

Adams v. Milwaukee, 228 U.S. 578.

Administrative Procedure Act, Section 21, Parts 146a–146e. Federal Register 22 (February 9, 1957): 849–850.

Administrative Procedure Act, Title 21, § 3.25, Federal Register, 16 (August 15, 1951): 8270.

Biggar, H. Howard. "The Old and the New in Corn Culture." *Yearbook of the United States Department of Agriculture, 1918.* Washington, D.C.: Government Printing Office, 1919.

Brainerd, W. K., and W. L. Mallory. "Milk Standards: A Study of the Bacterial Count and Dairy Score Card in City Milk Inspection." *Virginia Polytechnical Institute Agricultural Experiment Station Bulletin.* Bulletin 194 (1911).

Brew, James D. "Milk Quality as Determined by Present Dairy Score Cards." *New York Agricultural Experiment Station Bulletin.* Bulletin 398 (1915).

Brodell, Albert, and Thomas Kuzelka. "Harvesting the Silage Crops." United States Department of Agriculture, Statistical Bulletin Number 128. Washington, D.C.: Government Printing Office, 1953.

Bureau of Sanitary Engineering, Wisconsin State Board of Health. *Progress Report of the State Committee on Water Pollution.* Madison: Wisconsin State Board of Health, 1931.

Chapin, R. C. *The Standard of Living among Workingmen's Families in New York City.* New York: Charities Publication Committee, 1909.

Dairy Cattle Housing Subcommittee of the North Central Regional Farm Buildings Committee. *Dairy Cattle Housing in the North Central States.* Bulletin 470. Madison, Wisc.: Agricultural Experiment Station, 1947.

Davidson, Donald. *How Manufacturing Co-ops Market Grade A Milk.* Circular 26. Washington, D.C.: United States Department of Agriculture, Farm Cooperative Service, 1960.

Dawson, J. R. "Ice-Well Refrigeration for Dairy Farms Works Well at Mandan, N. Dak." In *United States Department of Agriculture, Yearbook of Agriculture*, 1931. Washington, D.C.: Government Printing Office, 1931.

Evidence Taken before the Joint Committee on Tuberculin Test, 1911: Volume I. Springfield, Ill.: Illinois State Journal Company, 1912.

Fischer v. St. Louis. 194 U.S. 361 (1904).

Hastings, E. G. "What Has Been Done with the Tuberculin Test in Wisconsin." *University of Wisconsin Agricultural Experiment Station Bulletin*. Bulletin 243. (October 1914).

Humphrey, G. C. "Community Breeders' Associations for Dairy Cattle Improvement." *Wisconsin Agricultural Experiment Station Bulletin*. Number 189 (February 1910).

Hunziker, Otto F., H. C. Mills, and H. B. Switzer. "Cooling Cream on the Farm." *Purdue University Agricultural Experiment Station Bulletin*. Bulletin 88 (1916).

Knapp, Harold. *The Effects of Deposition Rate and Cumulative Soil Level on the Concentration of Strontium-90 in U.S. Milk*. Division of Biology and Medicine, United States Atomic Energy Commission, Division of Technical Information. Washington, D.C.: Government Printing Office, 1961.

Lane, C. B., and Ivan Weld. "A City Milk and Cream Contest as a Practical Method of Improving the Milk Supply." United States Department of Agriculture, Bureau of Animal Industry, Circular 117. Washington, D.C.: Government Printing Office, 1907.

Larrabee, E. S., and G. Wilster. "The Butter Industry of Oregon: A Study of Factors Relating to the Quality of Butter." *Oregon State Agricultural College Bulletin*. Bulletin 258 (December 1929).

Meigs, Edward, et al. "The Relationship of Machine Milking to the Incidence and Severity of Mastitis." *United States Department of Agriculture*. Technical Bulletin Number 992. Washington, D.C.: Government Printing Office, 1949.

Milk and Its Relation to Public Health. Hygienic Laboratory Bulletin 56, March 1909. Washington, D.C.: Government Printing Office, 1912.

More, Louis Bolard. *Wage-Earners' Budgets: A Study of Standards and Cost of Living in New York City*. New York: Henry Holt, 1907.

Mortensen, M., and J. B. Davidson. "Creamery Organization and Construction: Part 1." *Iowa State College of Agriculture and Mechanic Arts*. Bulletin 139 (1913).

Mortensen, M., and B. W. Hammer. "Lacto: A Frozen Dairy Product." *Iowa State College of Agriculture and Mechanic Arts*. Bulletin 140 (1913).

Perry, E. J., and J. W. Bartlett. "Artificial Insemination of Dairy Cows." *New Jersey Extension Service, College of Agriculture*. Extension Bulletin 284 (1955).

Price, Bruce. "Marketing Country Creamery Butter by a Cooperative Sales Agency." *University of Minnesota Agricultural Experiment Station Bulletin*. Bulletin 244 (March 1928).

Russell, H. L. "Bovine Tuberculosis and the Tuberculin Test." *University of Wisconsin Agricultural Experiment Station*. Bulletin 40 (July 1894).

Russell, Henry Luman. "The Spread of Tuberculosis." In *Report of the Live Stock Sanitary Board, 1904*. Madison, Wisc.: Democrat Printing Company, 1905.

Russell, Judith, and Renee Fantin. *Studies in Food Rationing*. War Administration: Office of Price Administration, General Publication 13. Washington, D.C.: Government Printing Office, 1947.

Stocker, Noel. *Progress in Farm-to-Plant Bulk Milk Handling*. Farm Cooperative Service Circular 8. Washington, D.C.: United States Department of Agriculture, 1954.

U.S. Bureau of the Census. *Thirteenth Census of the United States Taken in the Year 1910. Volume 5: Agriculture*. Washington, D.C.: Government Printing Office, 1913.

U.S. Congress. House. Committee on Agriculture. *Oleomargarine: Hearings on H.R. 2400*. 78th Cong., 1st sess., 1943.

U.S. Congress. Joint Committee on Atomic Energy. *The Nature of Radioactive Fallout and Its Effects on Man: Hearing before the Special Subcommittee on Radiation of the Joint Committee on Atomic Energy*. 85th Congress, 1st sess. May 27–June 7, 1957.

U.S. Congress, Senate. Subcommittee of the Senate Committee on Foreign Relations. *Control and Reduction of Armaments*. 84th Congress. 2d sess., 1956. Sen. Res. 93.

U.S. Congress. House. *Tax on Adulterated Butter*. HR 1427, 62d Cong., 3rd sess., January 31, 1913.

U.S. Department of Agriculture. *Agricultural Statistics*. Washington, D.C.: Government Printing Office, 1957.

———. *Agricultural Statistics*. Washington, D.C.: Government Printing Office, 1960.

———. *Agricultural Statistics*. Washington, D.C.: Government Printing Office, 1965.

———. *Annual Yearbook of Agriculture, 1927*. Washington, D.C.: Government Printing Office, 1927.

———. *Annual Yearbook of Agriculture, 1931*. Washington, D.C.: Government Printing Office, 1931.

———. *Annual Yearbook of Agriculture, 1932*. Washington, D.C.: Government Printing Office, 1932.

———. *Annual Yearbook of Agriculture, 1935*. Washington, D.C.: Government Printing Office, 1935.

———. *Annual Yearbook of Agriculture, 1936*. Washington, D.C.: Government Printing Office, 1936.

———. *Annual Yearbook of Agriculture, 1942*. Washington, D.C.: Government Printing Office, 1942

———. *Annual Yearbook of Agriculture, 1943–1947*. Washington, D.C.: Government Printing Office, 1947.

———. *A Place to Live: The Yearbook of Agriculture, 1963*. Washington, D.C.: United States Department of Agriculture, 1963.

U.S. Department of Agriculture, Bureau of Agricultural Economics. *Production of Manufactured Dairy Products, 1946*. Washington, D.C.: Government Printing Office, May 1948.

U.S. Department of Agriculture, Bureau of Animal Industry. *Special Report on Diseases of Cattle*. Washington, D.C.: Government Printing Office, 1904.

U.S. Department of Commerce, Bureau of the Census. *U.S. Census of Agriculture, 1959: General Report*, vol. 2, *Statistics by Subjects*. Washington, D.C.: Government Printing Office, 1960.

U.S. Department of Health, Education, and Welfare. "Full-Scale System for Removal of Radio-strontium from Fluid Milk." Proceedings of Seminar, February 24–25, 1965. Public Health Service Publication 999-RH-28. Washington, D.C.: Government Printing Office, 1967.

U.S. Food and Drug Administration. "Government Strengthens Pesticide Controls." *FDA Report on Enforcement and Compliance* (April 1966): 7–8.

———. "Statement for Implementation of Report on No Residue and Zero Tolerance." 31 Federal Register (Wednesday, April 13, 1966): 5723–5724.

———. "Tolerances and Exemptions from Tolerances for Pesticide Chemicals in or on Raw Agricultural Commodities: Food Additives." 30 Federal Register (1965): 14328–14329.

"What Makes the Market for Dairy Products?" Bulletin 477. North Central Regional Publication. Madison: University of Wisconsin Agricultural Experiment Station, 1948.

White, Frank, and Clyde I. Griffith. "Barns for Wisconsin Dairy Farms." *University of Wisconsin Agricultural Experiment Station Bulletin*. Bulletin 266 (1916).

Witzel, S. A. "Dairy Cattle Housing." *University of Wisconsin Experiment Station Bulletin*. Bulletin 470. Rev. 1949. Madison: University of Wisconsin-Madison, 1949.

Witzel, S. A., and E. E. Heizer. *Loose Housing or Stanchion Type Barns? Summary of a 10 Year Dairy Cattle Housing Experiment in Southern Wisconsin*. Bulletin 503. Madison: University of Wisconsin Experiment Station, 1953.

Woll, F. W., and R. T. Harris. "A Decade of Official Tests of Dairy Cows, 1899–1909." *University of Wisconsin Agricultural Experiment Station*. Bulletin 194 (1910).

Secondary Sources

BOOKS AND ARTICLES

Abrahamson, Shirley Schlanger. "Law and the Wisconsin Dairy Industry: Quality Control of Dairy Products, 1838–1929." Ph.D. diss., University of Wisconsin-Madison, 1962.

Anderson, J. L. *Industrializing the Corn Belt: Agriculture, Technology, and Environment, 1945–1972.* DeKalb: Northern Illinois University Press, 2009.

Anderson, Oscar Edward. *Refrigeration in America: A History of a New Technology and Its Impact.* Princeton, N.J.: Princeton University Press, 1953.

Apple, Rima. *Mothers and Medicine: A Social History of Infant Feeding, 1890–1950.* Madison: University of Wisconsin Press, 1987.

———. *Vitamania: Vitamins in American Culture.* New Brunswick, N.J.: Rutgers University Press, 1996.

Atkins, Peter. *Liquid Materialities: A History of Milk, Science, and the Law.* Burlington, Vt.: Ashgate, 2010.

Barron, Hal. *Mixed Harvest: The Second Great Transformation in the Rural North, 1870–1930.* Chapel Hill: University of North Carolina Press, 1997.

Bederman, Gail. *Manliness and Civilization: A Cultural History of Gender and Race, 1880–1917.* Chicago: University of Chicago Press, 1995.

Beeman, Randall. "'Chemivisions': The Forgotten Promises of the Chemurgy Movement." *Agricultural History* 68 (1994): 23–45.

Belasco, Warren. *Appetite for Change: How the Counterculture Took on the Food Industry.* 2nd ed. Ithaca: Cornell University Press, 2007.

Benedict, Murray Reed, and Oscar Stine. *The Agricultural Commodity Programs.* New York: Twentieth Century Fund, 1956.

Bentley, Amy. *Eating for Victory: Food Rationing and the Politics of Domesticity.* Urbana: University of Illinois Press, 1998.

Black, John D. *The Dairy Industry and the AAA.* Washington, D.C.: Brookings Institution Press, 1935.

Blanke, David. *Sowing the American Dream: How Consumer Culture Took Root in the Rural Midwest.* Athens: Ohio University Press, 2000.

Block, Daniel Ralston. "Development of Regional Institutions in Agriculture: The Chicago Milk Marketing Order." Ph.D. diss., University of California–Los Angeles, 1997.

Borth, Christy. *Pioneers of Plenty: The Story of Chemurgy.* Indianapolis, Ind.: Bobbs-Merrill, 1939.

Bosso, Christopher. *Pesticides and Politics: The Life Cycle of a Public Issue.* Pittsburgh: University of Pittsburgh Press, 1987.

Bowers, William. *The Country Life Movement in America, 1900–1920.* Port Washington, N.Y.: Kennikat Press, 1974.

Boyd, William. "Making Meat: Science, Technology, and American Poultry Production." *Technology and Culture* 42 (2001): 631–664.

Boyer, Paul. *By the Bomb's Early Light: American Thought and Culture at the Dawn of the Atomic Age.* New York: Pantheon, 1985.

———. *Fallout: A Historian Reflects on America's Half-Century Encounter with Nuclear Weapons.* Columbus: Ohio State University Press, 1998.

———. *Urban Masses and Moral Order in America, 1820–1920.* Cambridge: Harvard University Press, 1978.

Buhs, Joshua Blu. "Dead Cows on a Georgia Field: Mapping the Cultural Landscape of the Post–World War II American Pesticide Controversies." *Environmental History* 7 (2002): 99–121.

———. *Fire Ant Wars: Nature, Science, and Public Policy in Twentieth-Century America.* Chicago: University of Chicago Press, 2004.

Burns, Sarah. *Pastoral Inventions: Rural Life in Nineteenth-Century American Art and Culture.* Philadelphia: Temple University Press, 1989.

Carmichael, Donald. "Forty Years of Water Pollution Control in Wisconsin: A Case Study." *Wisconsin Law Review* 1967 (Spring 1967): 350–419.

Cohen, Lizabeth. *Consumer's Republic: The Politics of Mass Consumption in Postwar America.* New York: Knopf, 2003.

————. *Making a New Deal: Industrial Workers in Chicago, 1919–1939.* New York: Cambridge University Press, 1990.

Cooke, Kathy. "The Limits of Heredity: Nature and Nurture in American Eugenics before 1915." *Journal of the History of Biology* 31 (1998): 263–274.

Cowan, Ruth Schwartz. *More Work for Mother: Ironies of Household Technology from the Open Hearth to the Microwave.* New York: Basic, 1983.

Craddock, Susan. *City of Plagues: Disease, Poverty, and Deviance in San Francisco.* Minneapolis: University of Minnesota Press, 2000.

Cronon, William. *Nature's Metropolis: Chicago and the Great West.* New York: Norton, 1991.

————. "The Trouble with Wilderness; or, Getting Back to the Wrong Nature." In *Uncommon Ground: Rethinking the Human Place in Nature,* ed. William Cronon, 69–90. New York: Norton, 1990.

Curry, Lynne. *Modern Mothers in the Heartland: Gender, Health, and Progress in Illinois, 1900–1930.* Columbus: Ohio State University Press, 1999.

Danbom, David. *The Resisted Revolution: Urban America and the Industrialization of Agriculture, 1900–1930.* Ames: Iowa State University Press, 1979.

Daniel, Pete. *Toxic Drift: Pesticides and Health in the Post–World War II South.* Baton Rouge: Louisiana State University Press, 2005.

Deutsch, Tracey. *Building a Housewife's Paradise: Gender, Politics, and American Grocery Stores in the Twentieth Century.* Chapel Hill: University of North Carolina Press, 2010.

Divine, Robert. *Blowing on the Wind: The Nuclear Test Ban Debate, 1954–1960.* New York: Oxford University Press, 1978.

Duffy, John. *The Sanitarians: A History of American Public Health.* Urbana: University of Illinois Press, 1990.

Dunlap, Thomas. *DDT: Scientists, Citizens, and Public Policy.* Princeton, N.J.: Princeton University Press, 1981.

Dupuis, E. Melanie. *Nature's Perfect Food: How Milk Became America's Drink.* New York: New York University Press, 2002.

Durand, Loyal. "The Lower Peninsula of Michigan and the Western Michigan Dairy Region: A Segment of the American Dairy Region." *Economic Geography* 27 (1951): 163–183.

Effland, Anne. "'New Riches from the Soil': The Chemurgic Ideas of Wheeler McMillen." *Agricultural History* 69 (Spring 1995): 288–297.

Egan, Michael. *Barry Commoner and the Science of Survival: The Remaking of American Environmentalism.* Cambridge: MIT Press, 2007.

Ehrenreich, Barbara. *The Hearts of Men: American Dreams and Flight from Commitment.* New York: Doubleday, 1983.

Enstad, Nan. *Ladies of Labor, Girls of Adventure: Working Women, Popular Culture, and Labor Politics at the Turn of the Twentieth Century.* New York: Columbia University Press, 1999.

Farrell, James J. *The Spirit of the Sixties: The Making of Postwar Radicalism.* New York: Routledge, 1997.

Finlay, Mark. "The Failure of Chemurgy in the Depression-Era South." *Georgia Historical Quarterly* 81 (Spring 1997): 78–102.

Fitzgerald, Deborah. *The Business of Breeding: Hybrid Corn in Illinois, 1890–1940.* Ithaca: Cornell University Press, 1990.

————. *Every Farm a Factory: The Industrial Ideal in American Agriculture.* New Haven: Yale University Press, 2003.

————. "'Farmers Deskilled: Hybrid Corn and Farmers' Work." *Technology and Culture* 34 (1993): 324–343.

Flanagan, Maureen. *America Reformed: Progressives and Progressivisms, 1890s–1920s.* New York: Oxford University Press, 2007.

Frank, Dana. *Purchasing Power: Consumer Organizing, Gender, and the Seattle Labor Movement, 1919–1929.* New York: Cambridge University Press, 1994.

Freidberg, Susanne. *Fresh: A Perishable History.* Cambridge: Harvard University Press, 2009.

Garrety, Karin. "Social Worlds, Actor-Networks, and Controversy: The Case of Cholesterol, Dietary Fat, and Heart Disease." *Social Studies of Science* 27, no. 5 (October 1997): 727–773.

Glickman, Lawrence. *Buying Power: A History of Consumer Activism in America.* Chicago: University of Chicago Press, 2009.

Gottlieb, Robert. *Forcing the Spring: The Transformation of the American Environmental Movement.* Washington, D.C.: Island Press, 1993.

Hale, William J. *The Farm Chemurgic: Farmward the Star of Destiny Lights Our Way.* Boston: Stratford Company, 1934.

Hamilton, Shane. *Trucking Country: The Road to America's Wal-Mart Economy.* Princeton: Princeton University Press, 2008.

Hayden, Dolores. *Building Suburbia: Green Fields and Urban Growth, 1820–2000.* New York: Pantheon, 2003.

Hays, Samuel P. *Beauty, Health, and Permanence: Environmental Politics in the United States, 1955–1987.* New York: Cambridge University Press, 1987.

———. *Conservation and the Gospel of Efficiency.* Cambridge: Harvard University Press, 1959.

Hazlett, Maril. "Woman vs. Man vs. Bugs: Gender and Popular Ecology in Early Reactions to *Silent Spring.*" *Environmental History* 9 (October 2004): 701–729.

Henderson, Henry, and David Woolner, eds. *FDR and the Environment.* New York: Palgrave-Macmillan, 2005.

Henninger-Voss, Mary, ed. *Animals in Human Histories: The Mirror of Nature and Culture.* Rochester, N.Y.: University of Rochester Press, 2002.

Herman, H. A. *Improving Cattle by the Millions: NAAB and the Development and Worldwide Application of Artificial Insemination.* Columbia: University of Missouri Press, 1981.

Horowitz, Roger. *Putting Meat on the American Table: Taste, Technology, Transformation.* Baltimore: Johns Hopkins University Press, 2006.

Hoy, Suellen. *Chasing Dirt: The American Pursuit of Cleanliness.* New York: Oxford University Press, 1996.

Ihde, Aaron John, and Stanley Becker. "Conflict of Concepts in Early Vitamin Studies." *Journal of the History of Biology* 4, no. 1 (Spring 1971): 1–34.

Jackson, John Brinkerhoff. *Discovering the Vernacular Landscape.* New Haven: Yale University Press, 1986.

Jackson, Kenneth. *Crabgrass Frontier: The Suburbanization of the United States.* New York: Oxford University Press, 1985.

Jacobs, Meg. *Pocketbook Politics: Economic Citizenship in Twentieth-Century America.* Princeton: Princeton University Press, 2005.

Jacoby, Karl. *Crimes against Nature: Squatters, Poachers, Thieves, and the Hidden History of American Conservation.* Berkeley: University of California Press, 2001.

Jones, Lu Ann. *Mama Learned Us to Work: Farm Women in the New South.* Chapel Hill: University of North Carolina Press, 2002.

Jones, Susan. *Valuing Animals: Veterinarians and Their Patients in Modern America.* Baltimore: Johns Hopkins University Press, 2003.

Jordan, Philip D. *The People's Health: A History of Public Health in Minnesota to 1948.* St. Paul: Minnesota Historical Society, 1953.

Judd, Richard. *Common Lands, Common People: The Origins of Conservation in Northern New England.* Cambridge.: Harvard University Press, 1997.

Keillor, Steven. *Cooperative Commonwealth: Co-ops in Rural Minnesota, 1859–1939.* St. Paul: Minnesota State Historical Society Press, 2000.

Kimmelman, Barbara. "The American Breeders' Association: Genetics and Eugenics in an Agricultural Context, 1903–1913." *Social Studies of Science* 13 (1983): 163–204.

Kirsch, Scott. "Harold Knapp and the Geography of Normal Controversy: Radioiodine in the Historic Environment." *OSIRIS* 19 (2004): 167–181.

Klaus, Alisa. *Every Child a Lion: The Origins of Maternal and Infant Policy in the United States and France, 1890–1920.* Ithaca: Cornell University Press, 1993.

Kline, Ronald. *Consumers in the Country: Technology and Social Change in Rural America*. Baltimore: Johns Hopkins University Press, 2000.

Klingle, Matthew. "Spaces of Consumption in Environmental History." *History and Theory* 42 (2003): 94–110.

Kloppenburg, Jack Ralph, Jr. *First the Seed: The Political Economy of Plant Biotechnology, 1492–2000*. New York: Cambridge University Press, 1988.

Kopp, Carolyn. "The Origins of the American Scientific Debate over Fallout Hazards." *Social Studies of Science* 9 (November 1979): 403–422.

Lampard, Eric. *Rise of the Dairy Industry in Wisconsin*. Madison: State Historical Society of Wisconsin, 1963.

Langston, Nancy. *Toxic Bodies: Hormone Disruptors and the Legacy of DES*. New Haven: Yale University Press, 2010.

Lanier, Raymond. "The Development of a Specialized Dairy Industry in Montana, 1919–1939." Master's thesis, Montana State University, 1956.

Lears, T. J. Jackson. *Fables of Abundance: A Cultural History of Advertising in America*. New York: Basic, 1994.

———. *No Place of Grace: Antimodernism and the Transformation of American Culture, 1880–1920*. New York: Pantheon, 1981.

Leavitt, Judith Walzer. *The Healthiest City: Milwaukee and the Politics of Health Reform*. Princeton: Princeton University Press, 1982.

Lehman, Tim. *Public Values, Private Lands: Farmland Preservation Policy, 1933–1985*. Chapel Hill: University of North Carolina Press, 1995.

Levenstein, Harvey. *Paradox of Plenty: A Social History of Eating in Modern America*. Berkeley: University of California Press, 2003.

———. *Revolution at the Table: The Transformation of the American Diet*. Berkeley: University of California Press, 2003.

Ludmerer, Kenneth. *Genetics and American Society: A Historical Appraisal*. Baltimore: Johns Hopkins University Press, 1972.

Lutts, Ralph. "Chemical Fallout: Rachel Carson's *Silent Spring*, Radioactive Fallout, and the Environmental Movement." *Environmental Review* 9 (1985): 210–215.

Maher, Neil. "A New Deal Body Politic: Landscape, Labor, and the Civilian Conservation Corps." *Environmental History* 7, no. 3 (July 2002): 435–461.

Manchester, Alden, and Don Blayney. *The Structure of Dairy Markets Past, Present, and Future*. Commercial Agriculture Division, Economic Research Service, USDA. Agricultural Economic Report No. 757. Washington, D.C.: Government Printing Office, 1997.

Marcus, Alan. *Agricultural Science and the Quest for Legitimacy*. Ames: Iowa State University Press, 1985.

Mason, Jennifer. *Civilized Creatures: Urban Animals, Sentimental Culture, and American Literature, 1850–1900*. Baltimore: Johns Hopkins University Press, 2005.

May, Elaine Tyler. *Homeward Bound: American Families in the Cold War Era*. Rev. ed. New York: Basic, 1999.

Mayo, James. *The American Grocery Store: The Business Evolution of an Architectural Space*. Westport, Conn.: Greenwood, 1993.

McCollum, Elmer V. *A History of Nutrition: The Sequence of Ideas in Nutrition Investigations*. Boston: Houghton Mifflin, 1957.

McMillen, Wheeler. *New Riches from the Soil: The Progress of Chemurgy*. New York: D. Van Nostrand, 1946.

McMurry, Sally. *Transforming Rural Life: Dairying Families and Agricultural Changes, 1820–1885*. Baltimore: Johns Hopkins University Press, 1995.

Meckel, Richard. *Save the Babies: American Public Health Reform and the Prevention of Infant Mortality, 1850–1929*. Baltimore: Johns Hopkins University Press, 1990.

Merchant, Carolyn. *Ecological Revolutions: Nature, Gender, and Science in New England*. Chapel Hill: University of North Carolina Press, 1989.

Miller, Julie. "To Stop the Slaughter of the Babies: Nathan Straus and the Drive for Pasteurized Milk, 1893–1920." *New York History* 74 (April 1993): 158–184.

Mink, Nicolaas. "It Begins in the Belly." *Environmental History* 14 (April 2009): 312–322.

Mintz, Sydney. *Sweetness and Power: The Place of Sugar in Modern History.* New York: Viking, 1985.

Mitman, Gregg. "In Search of Health: Landscape and Disease in American Environmental History." *Environmental History* 10 (2005): 184–210.

Mitman, Gregg, Michelle Murphy, and Christopher Sellers, eds. *Landscapes of Exposure: Knowledge and Illness in Modern Environments.* Chicago: University of Chicago Press, 2004.

Mooney, Patrick, and Theo Majka. *Farmers' and Farm Workers' Movements: Social Protest in American Agriculture.* New York: Twayne, 1995.

Muhm, Don. *The NFO: A Farm Belt Rebel.* Rochester, Minn.: Lone Oak Press, 2000.

Nash, Linda. *Inescapable Ecologies: A History of Environment, Disease, and Knowledge.* Berkeley: University of California Press, 2006.

Nash, Roderick. *Wilderness and the American Mind.* 3rd ed. New Haven: Yale University Press, 1982.

North, Charles E. "Milk and Its Relation to Public Health." In *A Half-Century of Public Health,* ed. Mazcyck Ravenel, 237–289. New York: American Public Health Association, 1921.

Oldenziel, Ruth, Adri Albert de la Bruheze, and Onno de Wit. "Europe's Mediation Junction: Technology and Consumer Society in the 20th Century." *History and Technology* 21 (March 2005): 107–139.

Olmstead, Alan, and Paul Rhode. "The 'Tuberculous Cattle Trust': Disease Contagion in an Era of Regulatory Uncertainty." *Journal of Economic History* 64 (2004): 929–963.

Opie, John. *Ogallala: Water for a Dry Land.* 2nd ed. Lincoln: University of Nebraska Press, 2000.

Paehlke, Robert. *Environmentalism and the Future of Progressive Politics.* New Haven: Yale University Press, 1989.

Paul, Oglesby. *Take Heart: The Life and Prescription for Living of Dr. Paul Dudley White.* Cambridge: Harvard University Press, 1986.

Pegram, Thomas. "Public Health and Progressive Dairying in Illinois." *Agricultural History* 65 (1991): 36–50.

Pernick, Martin. *The Black Stork: Eugenics and the Death of "Defective" Babies in American Medicine and Motion Pictures since 1915.* New York: Oxford University Press, 1996.

Phillips, Sarah Phillips. *This Land, This Nation: Conservation, Rural America, and the New Deal.* New York: Cambridge University Press, 2007.

Poppendieck, Janet. *Breadlines Knee-Deep in Wheat: Food Assistance in the Great Depression.* New Brunswick, N.J.: Rutgers University Press, 1986.

Price, Jennifer. *Flight Maps: Adventures with Nature in Modern America.* New York: Basic, 1999.

Pursell, Caroll, Jr. "The Farm Chemurgic Council and the United States Department of Agriculture, 1935–1939." *Isis* 60 (Autumn 1969): 307–317.

Rome, Adam. *The Bulldozer in the Countryside: Suburban Sprawl and the Rise of American Environmentalism.* New York: Cambridge University Press, 2001.

———. "'Give Earth a Chance': The Environmental Movement and the Sixties." *Journal of American History* 90 (September 2003): 525–554.

Rosenberg, Charles. *No Other Gods: On Science and American Social Thought.* Rev. ed. Baltimore: Johns Hopkins University Press, 1997.

———. "Pathologies of Progress: The Idea of Civilization as Risk." *Bulletin of the History of Medicine* 72, no. 4 (1998): 714–730.

Rosenkrantz, Barbara Gutman. "The Trouble with Bovine Tuberculosis." *Bulletin of the History of Medicine* 59 (1985): 155–175.

Rothman, Sheila. *Living in the Shadow of Death: Tuberculosis and Social Experience of Illness in American History.* Baltimore: Johns Hopkins University Press, 1994.

Ruble, Kenneth. *Men to Remember: How 100,000 Neighbors Made History.* Chicago: Lakeside Press, 1947.

Russell, Edmund. *War and Nature: Fighting Humans and Insects with Chemicals from World War I to Silent Spring*. New York: Cambridge University Press, 2001.

Ruttan, Vernon, ed. *Agricultural Policy in an Affluent Society*. New York: Norton, 1969.

Rydell, Robert. *All the World's a Fair: Visions of Empire at American International Expositions, 1876–1916*. Chicago: University of Chicago Press, 1984.

Sackman, Douglas Cazaux. *A Companion to American Environmental History*. Malden, Mass.: Wiley-Blackwell, 2010.

———. "Consumption and the Angel of History." *Environmental History* 10 (2005): 87–88.

———. *Orange Empire: California and the Fruits of Eden*. Berkeley: University of California Press, 2005.

Saloutos, Theodore. *The American Farmer and the New Deal*. Ames: Iowa State University Press, 1982.

Schrepfer, Susan, and Philip Scranton, eds. *Industrializing Organisms: Introducing Evolutionary History*. New York: Routledge, 1994.

Scott, Roy V. *The Reluctant Farmer: The Rise of Agricultural Extension to 1914*. Urbana: University of Illinois Press, 1970.

Seid, Roberta. *Never Too Thin: Why Women Are at War with Their Bodies*. New York: Prentice Hall, 1989.

Sellers, Christopher. *Hazards of the Job: From Industrial Disease to Environmental Health Science*. Chapel Hill: University of North Carolina Press, 1997.

———. "Thoreau's Body: Towards an Embodied Environmental History." *Environmental History* 4 (October 1999): 486–514.

Smith, Michael L., and Roland Marchand. "Corporate Science on Display." In *Scientific Authority and Twentieth-Century America*, ed. Ronald Walters, 148–167. Baltimore: Johns Hopkins University Press, 1997.

Stradling, David. *Conservation in the Progressive Era: Classic Texts*. Seattle: University of Washington Press, 2004.

———. *Smokestacks and Progressives: Environmentalists, Engineers, and Air Quality in America, 1881–1951*. Baltimore: Johns Hopkins University Press, 1999.

Strasser, Susan. "Making Consumption Conspicuous: Transgressive Topics Go Mainstream." *Technology and Culture* 43 (2002): 744–770.

———. *Waste and Want: A Social History of Trash*. New York: Metropolitan, 1999.

Sullivan, William Cuyler, Jr. *Nuclear Democracy: A History of the Greater Saint Louis Citizens' Committee for Nuclear Information, 1957–1967*. St. Louis, Mo.: Washington University Press, 1982.

Susman, Warren. *Culture as History: The Transformation of American Society in the Twentieth Century*. New York: Pantheon, 1984.

Swerdlow, Amy. *Women Strike for Peace: Traditional Motherhood and Radical Politics in the 1960s*. Chicago: University of Chicago Press, 1993.

Teller, Michael. *The Tuberculosis Movement: A Public Health Campaign in the Progressive Era*. Westport, Conn.: Greenwood, 1988.

Thomas, Roy, P. M. Reaves. and C. W. Pegram. *Dairy Farming in the South*. Danville, Ill.: Interstate Press, 1944.

Tomes, Nancy. *The Gospel of Germs: Men, Women, and the Microbe in American Life*. Cambridge: Harvard University Press, 1998.

Tucker, Richard. *Insatiable Appetite: The United States and the Ecological Degradation of the Tropical World*. Berkeley: University of California Press, 2000.

Valenčius, Conevery Bolton. *The Health of the Country: How American Settlers Understood Themselves and Their Land*. New York: Basic, 2002.

Valenze, Deborah. *Milk: A Local and Global History*. New Haven: Yale University Press, 2011.

Vilesis, Ann. *Kitchen Literacy: How We Lost Knowledge of Where Our Food Comes From and Why We Need to Get It Back*. Washington, D.C.: Island Press, 2008.

Wainwright, Milton. *Miracle Cure: The Story of Penicillin and the Golden Age of Antibiotics*. Oxford: Basil Blackwell, 1990.

Walker, J. Samuel. *Permissible Dose: A History of Radiation Protection in the Twentieth Century.* Berkeley: University of California Press, 2000.

Ward, Barbara Mclean, ed. *Produce and Conserve, Share and Play Square: The Grocer and the Consumer on the Home-Front Battlefield during World War II.* Portsmouth, N.H.: Strawbery Banke Museum, 1994.

Ward, John W., and Christian Warren, eds. *Silent Victories: The History and Practice of Public Health in Twentieth-Century America.* New York: Oxford University Press, 2007.

Warren, Louis. *The Hunter's Game: Poachers and Conservationists in Twentieth-Century America.* New Haven: Yale University Press, 1997.

Waserman, Manfred. "Henry L. Coit and the Certified Milk Movement in the Development of Modern Pediatrics." *Bulletin of the History of Medicine* 46 (1972): 359–390.

White, Richard. "Are You an Environmentalist or Do You Work for a Living? Work and Nature." In *Uncommon Ground: Rethinking the Human Place in Nature*, ed. William Cronon, 171–185. New York: Norton, 1995.

Whorton, James. "'Antibiotic Abandon': The Resurgence of Therapeutic Rationalism." In *History of Antibiotics: A Symposium*, ed. John Parascandola, 125–136. Madison, Wisc.: American Institute for the History of Pharmacy, 1980.

Williams, Raymond. *The Country and the City.* New York: Oxford University Press, 1975.

Wilson, Alexander. *The Culture of Nature: North American Landscape from Disney to the Exxon Valdez.* Malden, Mass.: Blackwell, 1992.

Wolf, Jacqueline. *Public Health and the Decline of Breastfeeding in the 19th and 20th Centuries.* Columbus: Ohio State University Press, 2001.

Wood, Andrew. *New York's 1939–1940 World's Fair.* Charleston, S.C.: Arcadia, 2004.

Worster, Donald. *The Dust Bowl: The Southern Plains in the 1930s.* Oxford: Oxford University Press, 1979.

Wright, David. "Agricultural Editors Wheeler McMillen and Clifford V. Gregory and the Farm Chemurgic Movement." *Agricultural History* 69 (Spring 1995): 272–287.

———. "Alcohol Wrecks a Marriage: The Farm Chemurgic Movement and the USDA in the Alcohol Fuels Campaign in the Spring of 1933." *Agricultural History* 67 (Winter 1993): 36–66.

INDEX